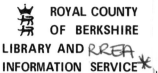

THE STUFF OF THE UNIVERSE

THE STUFF
OF THE UNIVERSE

Dark Matter,
Mankind and the
Coincidences of Cosmology

JOHN GRIBBIN
AND MARTIN REES

HEINEMANN : LONDON

William Heinemann Ltd
Michelin House, 81 Fulham Road, London SW3 6RB

LONDON   MELBOURNE   AUCKLAND

First published in Great Britain 1990
Copyright © 1989 by John Gribbin and Martin Rees
Reprinted 1990 (twice)

A CIP catalogue record for this book
is available from the British Library

ISBN 0 434 62636 8

Printed in Great Britain
by St Edmundsbury Press Ltd, Bury St Edmunds, Suffolk
and bound by Hunter & Foulis Ltd, Edinburgh

CONTENTS

---★---

PART THREE:
The Bespoke Universe
239

INTRODUCTION

★

Why Are We Here?

THERE ARE THREE MOTIVES for studying the Universe. The first is discovery: to learn what's out there, whether in our own Solar System or in the extragalactic realm. This vicarious exploration—of the surface of Mars, or the patterns of spiral galaxies—is something a wide public can share.

For the astrophysicist, this exploration is preliminary to a second goal: to understand and interpret what we see, in terms of the laws of physics established here on Earth, and to place our entire Solar System in an evolutionary context that can be traced back to the birth of the Milky Way Galaxy, and beyond—right back, indeed, to the initial instants of the so-called Big Bang with which our Universe began.

To the physicist, there is a third motive: The cosmos is a "laboratory" offering more extreme conditions than can be simulated on Earth. Known laws can be tested, perhaps to the breaking point, by applying them, for instance, to the amazing densities of neutron stars; and a better understanding of the astounding temperatures and energies of the Big Bang could reveal new laws. Essentially *all* that we know about gravity—one of the four fundamental forces, and the one that controls the motions of stars, galaxies, and the entire expanding Universe—comes from astronomy.

Astronomy is, of course, an old pursuit—perhaps it was the first science to become professionalised—but it has greatly enlarged its scope during the past two decades. Recent progress has been largely "driven" by experimental and observational advances. No armchair theorist, even equipped with current physical knowledge, could have envisaged the extraordinary phenomena and objects that have been discovered. This burgeoning is due partly to technical improvements in optical astronomy, but even more to the new windows on the Universe opened up by radio astronomy and by observations from space. Valuable data are also obtained in other ways—from underground neutrino detectors and gravitational-wave experiments. There are few branches of terrestrial physics, indeed, that do not find application somewhere in astronomy.

In this book, we have (especially in the middle section) described those recent developments that we have found (from our experience of lecturing and writing) that seem to fascinate nonspecialists most. We aim to answer the questions that we most often are asked. Few of these topics—quasar spectra, protogalaxies, gravitational lenses, gravitational waves, and cosmic strings—have yet been given due prominence in nontechnical publications. On the other hand, stories such as that of black holes are not emphasised here because such exotic objects have become so familiar from the many excellent books that already exist.

All these topics relate to a single overall conclusion—something that has as much right to be called a paradigm shift as anything in twentieth-century astronomy. This is the realisation that the dynamics of our Universe, and of all the galaxies in it, are controlled not by what we see but by *dark matter*. Only 10 percent (at most) of the Universe shines; what we see is a biased and incomplete sample of the Universe's overall contents. Without the dark matter, our Universe would be a very different place: Dark matter controls the structure and eventual fate of the Universe. Discovering what the "dark stuff" is surely rates as the number-one problem confronting cosmologists today.

The search for a solution to this puzzle is a natural development from recent discoveries in cosmology that have been reported in earlier books. A fuller description of Big Bang cosmology and the expanding Universe can be found in *In Search of the Big Bang;* the ultimate fate of the Universe, and evidence that dark matter does indeed exist, are discussed in detail in *The Omega Point*. Here, moving on from such discoveries, we are more concerned with the exact nature of the dark matter, the stuff of the Universe, than with the detailed proof that there is some sort of dark matter around.

It is no exaggeration to say that we would not be here to wonder at the Universe if the dark stuff were *not* around. We can imagine ways in which the Universe might have emerged from the Big Bang without this background sea of stuff, so that stars, galaxies, and creatures like us would never have been produced. And yet we are here, and this relates to the second main theme of our book.

Science deals mostly with complex manifestations of laws that in essence are well known—the real scientific challenge lies in understanding the rich complexity inherent in these phenomena. Cosmology and particle physics are, however, the two frontier areas, where even the basic laws are still mysterious. Moreover, deep interconnections are becoming apparent between these two endeavours—the study of the cosmos and of the microworld. For example, the dark matter that dominates the Universe is probably in the form of myriads of tiny particles whose individual properties can be understood only in microphysical terms.

The study of the Universe, and our place in it, evolves in a piecemeal way. We can make progress only by tackling problems in "bite-sized" pieces, and specialists are perforce concerned with technical details. But the occupational risk of astrophysicists, and indeed of all scientists, is to forget that one is wearing blinkers—that there are broader questions at issue, and that a main goal of our piecemeal efforts is eventually to elucidate them.

Why is our Universe the way it is? What is our place in it? Could things have been otherwise, and could alternative universes exist? Why, above all, does the Universe have the symmetry and simplicity that have allowed us to make any progress in understanding it? These issues, where even the specialists are still groping for clues, are the ones that come up most frequently in general discussions. Their investigation sometimes goes by the name of *anthropic cosmology*—but giving the investigation a name doesn't mean that we yet have all, or any, of the answers.

As the frontier of cosmological knowledge has advanced, its periphery has expanded, and issues that were once purely conjectural have come within the scope of serious investigation. Questions about how the Universe began and how it may end can now be addressed scientifically, and not just in our nonprofessional moments. Such debates have been the subjects of previous books; here, we have not shied away from more speculative issues, coming just within the periphery of respectable science, and we try to give the flavour of current debates that are on (but not beyond) the frontiers of the subject.

The unifying theme of this book can be stated in nontechnical terms—indeed, there is no other way to express it, since the specialists are as perplexed about the answer as anyone else: *What features of the Universe were essential for the emergence of creatures such as ourselves, and is it through coincidence, or for some deeper reason, that our Universe has these features?* We hope our discussion of these issues will answer some of the questions in your mind.

John Gribbin
Martin Rees
October 1988

PART ONE

★

Cosmic Coincidences

CHAPTER ONE

————————————★————————————

How Special Is the Universe?

SCIENCE IS NOT SIMPLY the accumulation of more and more facts about the natural world. If that were the case, science would long since have ground to a halt, clogged up by the accumulation of vast amounts of data. Instead, science proceeds because of our ability to discern patterns and regularities in the natural world. As we come to see how previously unconnected facts hang together, we fit more data into laws of greater scope and generality, and we need to remember *fewer* independent basic facts, from which all the rest can be deduced. The astonishing triumph of modern science, especially physics and astronomy, is its ability to describe so many of the bewildering complexities of the natural world in terms of a few underlying principles. But this success seems to rest upon the fact that our Universe is "constructed" along very simple lines. The laws of physics are straightforward enough to be understood by human minds, and the laws we deduce from experiments here on Earth seem to apply across the Universe, at all places and at all times. Is such simplicity an inevitable feature of the Universe? Is it

merely a coincidence that creatures intelligent enough
to understand a few simple physical laws exist in a
world where only those physical laws are needed to
explain how everything works? Or is there some deeper
plan that ensures that the Universe is tailor-made for
humankind?

These questions, which relate to our place in the
Universe, and are concerned with the issues of what
has been dubbed *anthropic cosmology*, are addressed in
this book. The success of science in explaining complex
patterns of behaviour by simple laws can be seen by a
few examples. The regular courses of the Moon and
planets across the sky had been known since ancient
times, but were explained only when Newton realised
that they were governed by the same gravitational force
that holds us down on the Earth. And the complexity of
chemistry, which so baffled the alchemists, began to be
understood when Mendeleev, in the nineteenth century,
found regularities in the way the properties of elements
related to one another; these regularities are now attributed to the fact that atoms are made from just
three basic types of component, the protons and neutrons (together making up the nucleus) and the electrons (which are distributed outside the nucleus in
accordance with the laws of quantum mechanics).

Physicists have now reduced nature still further. They
believe that the basic structure of the entire physical
world—not just atoms but stars and people as well—is
in principle determined by a few basic "constants."
These are the masses of a few so-called elementary
particles, and the strengths of the forces—electric, nuclear, and gravitational—that bind those particles together and govern their motions.

In terms of these simple rules, some natural phenomena are more easily explained than others. Biological
processes, for example, are much harder to understand
than the fall of an apple from a tree or the orbit of a

planet around the Sun. But it is *complexity*, not sheer size, that makes a process hard to comprehend. We already understand the inside of the Sun better than we do the interior of the Earth. The Earth is harder to understand because the temperatures and pressures inside it are less extreme, and therefore more subtle, than those inside the Sun. Complex structures—chemical compounds containing many atoms joined together—exist inside the Earth; inside the Sun, however, everything is reduced by the heat and pressure to the constituent atomic nuclei and electrons, and their behaviour is governed by the basic rules.

Our Universe contains thousands of millions of galaxies, and each of those galaxies may, like our Milky Way, contain thousands of millions of stars, more or less like our Sun. Observations show that the Universe is expanding, with groups of galaxies moving apart from one another as time goes by. Cosmologists infer that there was a time, roughly 15 billion years ago, when all of the matter and energy of the Universe, and space and time as well, were concentrated in a superhot region, a fireball known as the Big Bang. In the earliest stages of the primordial fireball, matter would surely have been reduced, or broken down, into its most primitive constituents—a "thermal soup" at a temperature of 10 billion degrees Celsius, initially expanding at such a rate that it doubled in size every second. In this sense, conditions in the Big Bang were even simpler than those inside the Sun today. So we can realistically hope to explain why the Universe is expanding the way it is—it isn't presumptuous to try to understand the physics involved. Perhaps we can also understand how stars and galaxies came into existence in the expanding Universe, and therefore begin to appreciate the nature of our own origins. But as soon as we begin to gain an understanding of these processes, we immediately run into the puzzle of the cosmic coincidences.

The Anthropic Universe

The Universe is a simple place, but we are complex creatures. One reason for this is that we do not inhabit a typical place in the Universe. Most of the Universe is empty space, filled with a weak background sea of electromagnetic radiation, with a temperature only 3 degrees above the absolute zero of temperature, which lies at −273 degrees C. But we live on a planet, which orbits around a simple, stable star. Conditions inside that star—our Sun—provide the energy that life, including human life, needs; conditions on the surface of that planet—the Earth—allow for the complexity that seems essential to life as we know it. Clearly, our home represents a special place in the Universe (although not necessarily a *unique* place). Slightly more subtly, we can see that we also exist at a special *time* in the Universe. In the Big Bang itself, conditions were too extreme for the complexity that represents human life to exist; today, they are just right (at least on one planet, orbiting one star in one galaxy). In the future, perhaps conditions will once again be unsuitable for life as we know it. We exist here and now because of the exact relationships between the basic forces and particles. And this raises many questions.

Why, for example, are stars so big? The strength of the electrical force between two protons (in, say, a hydrogen molecule) can be compared with the gravitational force between the same two particles. Electrical forces are 10^{36} (a 1 followed by 36 zeroes) times stronger than gravitational forces, and on the scale of an atom gravity can be completely ignored. But when large numbers of atoms are grouped together, the force of gravity increases as the total mass increases. Each atom has zero net electrical charge, because the positive charge on each proton is exactly balanced by the negative charge on an electron in the atom (some people, incidentally, see this exact balance between the charge on an electron and the charge on a proton as a remarkable coincidence in its own right). So, a large mass carries no net electrical charge and exerts no net electrical force. When an apple falls from a tree, it does so not because of the electrical forces pulling it towards the Earth, but because of the accumulated gravitational force of the enormous numbers of atoms that together make up the Earth. In fact, the apple is held together by electrical forces, acting between its constituent atoms and molecules. The same forces hold together the atoms and molecules of the stem that attaches the apple to the tree. The apple falls if, and when, the gravity of the whole Earth is strong enough to overcome the electrical forces in the stem and break the apple free from its parent tree. The gravity of the whole Earth is needed to break the electrical forces involving the relatively few atoms in the stem of the apple.

Theoretical studies of stars and their life cycles were stimulated by the challenge of observations—people saw the stars and wondered what they were made of. It is interesting, though, that the properties of stars could have been deduced by a physicist who lived on a perpetually cloud-bound planet. Such a physicist could have posed the question: Can one have a gravitationally bound fusion reactor, and what would it be like? He or she might then have reasoned like this: Because gravity is never cancelled out in the way electrical charges cancel, it must win out over electrical forces on a sufficiently large scale. But how large?

Imagine that we assemble a set of objects containing successively 10, 100, 1,000 atoms, and so on. The 24th object would be the size of a sugar lump—about 1 cubic centimetre. The 39th would be like a rock 1 kilometre across. Gravity starts off with a "handicap" of 10^{36}, but it gains on electrical forces as a two-thirds power.* So when we get to our 54th object, because 36 is two-thirds of 54, it will have caught up. Our 54th object will have the mass of Jupiter; anything bigger than Jupiter will start to get crushed by gravity. So, to be squeezed by gravity and heated to the point where nuclear fusion could ignite, an object must contain well over 10^{54} atoms.

Gravitationally bound fusion reactors—stars—must be massive because gravity is so weak. Having inferred this, our hypothetical physicist could in principle calculate the entire life cycle of a star. Sir Arthur Eddington was the first person to express this line of argument clearly, in the 1920s; he went on to conclude that "when we draw aside the veil of clouds beneath which our physicist is working and let him look up at the sky, there he will find a thousand million globes of gas, nearly all with masses [in this calculated range]."

Gravity dominates the electrical forces, and crushes atoms out of existence, when the total mass of a collection of atoms approaches 10^{57} times the mass of a proton. Even the interior of the Earth can resist the inward pressure of gravity and maintain atoms as distinct entities. But when the total mass nears that critical value, the structure of atoms is destroyed. What remains is a sea of freely mingling nuclei and electrons. Stars do indeed have masses around 10^{57} times the mass of a proton. They are held together by gravity, and gravity initiates the process of nuclear fusion, when atomic nuclei are squeezed together to make new nuclei, which provides the energy that keeps stars hot. If gravity were even weaker, stars would be bigger still; if gravity were stronger, stars would be smaller and would

*The reason is simple: the force involved depends on mass M and radius R and is proportional to M/R; for uniform density, mass is proportional to the cube of radius, that is, radius is the one-third power of the mass, and M/R goes as the two-thirds power of mass.

run through their life cycles more quickly—perhaps so quickly that there would be no time for intelligent life to evolve on any planets orbiting those stars.

The basic forces also determine how big a human being can be. Our bodies, like all chemical structures, are held together by electrical forces. These forces are fixed by the basic laws of nature. But because the gravitational force acting on our bodies—our weight— depends on how many atoms the bodies contain, the force is bigger if people (or other creatures) are bigger. The bigger they come, the harder they fall. A simple calculation shows that any creatures much bigger than a human being, inhabiting the surface of a planet the size of the Earth, will simply break apart when they fall over. We are as big as we can be, given our lifestyle—or rather, the lifestyle of our recent ancestors. Whales can be big, because their mass is supported by the sea; but our ancestors, who were tree-dwelling primates, couldn't be so big that an occasional fall would inevitably prove fatal.

We shall look in more detail at these, and other cosmic coincidences in part 3 of the book. But it is worth spelling out now just how delicate the balances between the basic forces that permit our existence really are. For example, if the nuclear forces, which control the behaviour of protons and neutrons within the nucleus of an atom, were slightly stronger than they actually are, compared with electrical forces, then the di-proton (an atomic nucleus composed of two protons) would be stable. In our Universe, the electrical force of repulsion between two positively charged protons overwhelms the nuclear force of attraction between them, and di-protons do not exist. Two protons can be held in a stable atomic nucleus only if there is a neutron or two there as well; these uncharged particles add to the attractive force but do not affect the repulsive force. Now, stars gain their energy by fusing protons and neutrons together into such nuclei; if, instead, they could fuse pairs of protons together into di-protons, stars would evolve quite differently and the Universe would be a very different place. If, on the other hand, nuclear forces were slightly weaker than they are in our Universe, no complex nuclei could form at all. The entire Universe would be composed of hydrogen, the simplest element, whose atoms consist of a single proton and a single electron.

All the familiar chemical elements except hydrogen and primordial helium were, in fact, built up by nuclear transmutations inside stars that exploded long before our Solar System formed. Iron, carbon, oxygen, and the rest are all products of stellar nucleosynthesis, a process that is sensitive to several apparent accidents of physics, as Fred Hoyle pointed out in the 1950s. We shall look in detail at those coincidences later; what matters here is that the Universe seems to have been set up in such a way that interesting things can happen in it. It is very easy to imagine other kinds of universes, which would have been stillborn because the laws of physics in them would not have allowed anything interesting to evolve.

Imagine, for instance, tinkering with the Universe by varying the strength of gravity. Suppose that it were only 10^{26}, rather than about 10^{36}, times weaker than the electrical force. We would then have a smaller Universe, in which stellar processes would occur more rapidly. Stars, which are fusion reactors bound together by gravity, would each have only about 10^{-15} (one-millionth of a billionth) of the Sun's mass. Although each one would have a mass of a trillion tons, it would take 10 million of them to add up to the mass of our Moon; and each would last for just about a year before burning out. Very probably, this would not provide time for life forms as complex as ourselves to evolve; in any case, complex structures could not grow very large before being crushed by gravity.

So the fact that we exist tells us, in a sense, what conditions are like inside stars and in the Universe at large. This is the mildest form of what is now known as anthropic reasoning, or anthropic cosmology. Given the brute fact that we are a carbon-based form of life, which evolved slowly on a planet orbiting around a star like our Sun (a so-called G-type star), there are some features of the Universe, some constraints on the possible values of physical constants, which can be inferred quite straightforwardly. This line of reasoning even helps us to understand the sheer size of the Universe.

A Universe Big Enough for Life

At first sight, it might seem that one planet like the Earth, circling one star like our Sun, would be sufficient to provide a home for life and the opportunity for intelligence to evolve. There is no way to set a precise figure on the extent of our Universe and the number of stars and planets it contains, but at the very least it contains a billion billion (10^{18}) stars, and at least 1 percent of that number—some 10 million billion stars—are likely to be reasonably similar to our Sun. If we guessed that just 1 percent of those Sun-like stars actually possessed a retinue of planets that included a planet like the Earth, that would still provide a hundred thousand billion homes for life as we know it. This is a number so extravagantly large that it makes our place in the Universe seem utterly insignificant. And yet it may be necessary that all those billions of potential homes for life exist, simply because *one* home for life, our Earth, exists.

Consider the implications in terms of the linear size of the Universe, rather than the number of stars it contains. Cosmologists estimate, for good, sound reasons,* that the observable Universe is about 15 billion light-years across. A light-year is simply the distance that light can travel in one year, so it is no coincidence that this size is linked to the estimated age of the Universe—15 billion years. We can, in principle, "see" as far as light has had time to travel since the Universe began.

The fireball of the Big Bang was a simple place, in the sense that matter was broken down into its component parts there. As the Universe expanded and cooled, those basic building blocks of matter formed into the simplest elements, hydrogen and helium. But studies of the light from very old stars show that scarcely any heavier elements than these emerged from the Big Bang. The essential molecules of life, which include carbon, oxygen, nitrogen, and phosphorus, were manufactured by thermonuclear processes inside stars *after* the Big Bang. Our own Sun is not one of the first stars that formed when the Universe was young. Those stars went through their life cycles, converting hydrogen and helium into more complex nuclei, and some of those stars then exploded as supernovae, scattering the fruits of stellar nucleosynthesis through the dust and gas clouds of the young Galaxy. Only later generations of stars, born out of collapsing fragments of those interstellar clouds, contained enough of the heavier elements to form planets like the Earth, and allowed life forms like ourselves to emerge.

All that took time. In round terms, it takes a few billion years for a galaxy to form, for the first stars in it to process hydrogen and helium into heavier elements,

---

*The details can be found in *In Search of the Big Bang*, by John Gribbin (see bibliography).

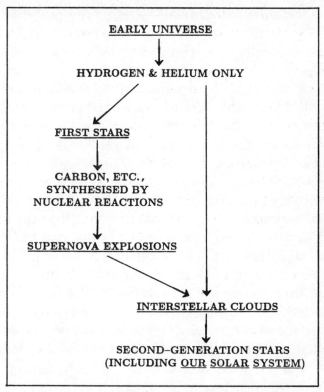

Figure 1.1 The history of the atoms on Earth.

live out their lives, and die in a blaze of glory, scattering those elements in the process. It then takes more time for new stars to form out of the debris, and for life to evolve on the planets circling those stars. In order for us to be here wondering about it all, the Universe *must* be about 15 billion years old, and therefore about 15 billion light-years across!

This insight demonstrates the power of anthropic reasoning. Simply from the fact that we are a carbon-based life form we can deduce that the Universe must be a certain size and a certain age. Sometimes, those who argue that the Universe cannot possibly have been designed, or created, expressly in order to produce a carbon-based, intelligent life form inhabiting a single planet orbiting an ordinary star, point out that the Universe seems ludicrously *over*designed for such a task bigger and older, and containing far more stars, than seems necessary. Provided that the laws of physics had to be as they are, the argument falls down. Given the laws of physics that operate in our Universe, all those billions of stars and billions of light-years are necessary for our existence. The argument returns in full force, however, if you assume that whoever designed the Universe could have chosen different values of the constants of nature.

It is, perhaps, also worth pointing out that the argument is not affected, one way or the other, if there are forms of intelligent life in the Universe that do not depend on carbon chemistry for their being. Science fiction authors, and some who write nonfiction, have speculated on the possibilities of, for example, life on the surface of a neutron star, or intelligence generated by magnetic fields that eddy through a black cloud in space. But *we* are a carbon-based life form, and therefore it is no surprise to find that we see a Universe 15 billion years old and 15 billion light-years across. We do *not* observe the Universe at a randomly chosen instant of cosmic time, but at just about the earliest time that life forms like us could begin to be asking questions about the Universe at large.

But still, there is something very curious about the fact that the Universe has expanded away from the Big Bang at just the right speed to allow galaxies, stars, and planets to form, and for carbon-based life forms to exist on at least one of those planets. It is a puzzle that is almost too obvious to seem worthy of attention— why is there anything interesting here at all?—but that points to the most astonishing cosmic coincidence of all. It has to do with how much matter the Universe contains, and how fast it expands: In more technical language, how "flat" the spacetime of the Universe must be.

The Primary Puzzle

One way to get a grip on the most extreme cosmic coincidence of all is by looking again at the variety of chemical elements in our Universe, and especially on our planet. The reason hydrogen and helium can be converted into heavier elements inside stars is that heavier elements, once formed, represent a more efficient way of storing matter. Protons and neutrons stored in the form of carbon nuclei are held together more effectively than similar particles in nuclei of helium. This is why converting helium nuclei into carbon nuclei releases energy, which helps to keep a star hot. Because all atomic nuclei carry a positive charge, proportional to the number of protons they contain, they must be pushed together hard enough to overcome the electrical repulsion force, and close enough to allow the nuclear force, which is stronger but has a shorter range than the electrical force, to dominate. Nuclear fusion won't happen, therefore, unless the nuclei have big random motions, leading to hard collisions—unless, in other words, the temperature is very high. If 10^{57} particles are gathered together in a star, gravity can confine and squeeze them to sufficiently high temperatures. Gravity overcomes the electrical force and allows the nuclear force to get to work.

Conditions in the Big Bang were also extreme enough to do the trick. It was hot, and the pressure was very high. At first, however, protons and neutrons, which had just been manufactured out of pure energy, could not stick together in complex nuclei because they would be torn apart by the battering of repeated collisions with other particles. But as the Universe expanded it started to cool—in exactly the same way as gas expanding out of a confined space cools, the principle on which a domestic refrigerator operates. There must have come a time when conditions were right for protons and neutrons to be welded together into nuclei of heavy elements. The process began with the production of helium nuclei, each containing two protons and two neutrons, and it would have proceeded quite quickly to heavier elements, if those conditions had persisted. The most energetically stable nucleus of all is that of iron, and if the Universe had cooled slowly enough, then most of the protons and neutrons would have been locked up in iron nuclei. Had that been the case, the Universe would have been a dull place in which no further interesting reactions could have occurred; stars would not exist, and there would have been no opportunity to build up the familiar biological complexity of life on Earth.

The crucial factor that prevented all the primordial matter from turning into iron, but allowed stars like our Sun to form, and to build up a variety of elements starting from hydrogen and helium, was the rate at which the early Universe expanded. The faster the expansion, the quicker the cooling; and the higher the density of nuclei, the more likely that reactions will go completely through to equilibrium within the time available. Analysis of the light from old stars shows that just 25 percent of the matter emerging from the Big Bang was in the form of helium, and virtually all the rest was still in the form of hydrogen. Hardly any nuclei more massive than helium formed in the Big Bang. This simple number, the ratio of hydrogen to helium in old stars, actually tells cosmologists a great deal about the content of the Universe when it was only one second old, and how quickly it was expanding and cooling during the Big Bang.

The early Universe must have been a mixture of nuclei and radiation, with the radiation component being overwhelmingly dominant. Calculations of nuclear reactions suggest that there was only 1 nucleus for every 2×10^9 quanta of radiation—or photons. This ratio has held throughout the subsequent life of the Universe. Today there are around 400 photons in every cubic centimetre of space, so the calculations imply a mean density of 1 atom in every 5 cubic metres of space. This is, indeed, roughly consistent with what astronomers see, if all the matter in all the stars were spread out evenly; we will put these numbers in their proper perspective later.

Are there other constraints on the rate at which the Universe could be "allowed" to expand, given that we exist? After the hydrogen and helium that emerged from the Big Bang had cooled, they began to form into clouds of gas that were held together by gravity. Some of those clouds collapsed, under the pull of their own gravity, even though the Universe as a whole was expanding. Embryonic galaxies could have been simply regions of the Universe where the density was slightly greater than the average, regions whose expansion therefore lagged behind the overall expansion of the Universe, forming collapsed clouds of gas that became stars and galaxies. This happened when the Universe was about 10 percent of its present age—between 1 and 2 billion years after the Big Bang. We don't know exactly how galaxies formed, but it is clear that if the Universe had been expanding too rapidly, the clouds would have been spread thin and pulled apart before gravity could dominate, even on a local scale, and make them collapse into galaxies and stars. Without that collapse, no heavy elements would have been cooked in stellar interiors and, once again, we would not be here to wonder about the nature of the Universe. On the other hand, if the Universe had started off expanding too slowly, it would have come to a halt and started to recollapse by now, with galaxies falling towards each other. We can even imagine a universe in which the expansion reversed within the first million years; incipient galaxies and stars would have been snuffed out before they could even have had a chance to form.

So our existence tells us that the Universe must have expanded, and be expanding, neither too fast nor too slow, but at just the "right" rate to allow elements to be cooked in stars.

This may not seem a particularly impressive insight. After all, perhaps there is a large range of expansion rates that qualify as "right" for stars like the Sun to exist. But when we convert the discussion into the proper description of the Universe, Einstein's mathematical description of space and time, and work backwards to see how critical the expansion rate must have been at the time of the Big Bang, we find that the Universe is balanced far more crucially than the metaphorical knife edge. If we push back to the earliest time at which our theories of physics can be thought to have any validity, the implication is that the relevant number, the so-called "density parameter," was set, in the beginning, with an accuracy of 1 part in 10^{60}. Changing that parameter, either way, by a fraction given by a decimal point followed by 60 zeroes and a 1, would have made the Universe unsuitable for life as we know it. The implications of this finest of finely tuned cosmic coincidences form the heart of this book, and in part 2 we shall discuss the strange forms of dark matter that may exist in the Universe today, and against which all the matter in all the bright stars of all the visible galaxies represents less than the tip of the proverbial iceberg. That dark matter may be as crucial to the existence of those stars as the plenitude of stars is to the existence of life on Earth.

The Flat Universe

Gravity is the controlling force of planets, stars, and all large astronomical systems, including the Universe itself. On Earth, and throughout the Solar System, Newton's theory of gravity, which says that the force of attraction acting between two bodies is inversely proportional to the square of their distance apart, is an excellent approximation (although some recent investigations suggest that Newton's law may need to be modified very slightly, even on Earth, by an effect known as the "fifth force"). But when gravity is stronger, which happens when objects are compressed into very small volumes or masses even larger than stars are involved, Newtonian ideas are inadequate to describe gravitational effects. The theory that goes beyond Newton to provide a working description of gravity under such extreme conditions is Einstein's theory—general relativity.

In academic libraries, the old back numbers of scientific journals are usually seldom consulted and are relegated to remote stockrooms. But two particular volumes—the issues of *Annalen der Physik* for 1905 and 1916—are treasured collectors' items: They contain the papers by Albert Einstein that established him as the greatest physicist since Newton.

In 1905, the 26-year-old Einstein not only elucidated his theory of "special relativity"; he also proposed that light is quantised into packets of energy (photons), and formulated the statistical theory of how tiny particles move through the air or a liquid (Brownian motion). These contributions alone rank him among the half-dozen great pioneers of twentieth-century physics.

But it is his gravitational theory, "general" relativity, developed ten years later, that puts Einstein in a class by himself. Even if he had contributed none of his 1905 papers, it would not have been long before the same concepts were put forward by some of his distin-

guished contemporaries. The ideas were "in the air";
well-known inconsistencies in earlier theories, and puzzling experimental results, were focussing attention on
these problems. But general relativity, the interpretation of gravity in terms of curved spacetime so that
"space tells matter how to move; matter tells space
how to curve," was not a response to any particular
observational enigma. True, it did account for an old
puzzle about the orbit of Mercury, and it was famously
confirmed by measurements made during an eclipse of
the Sun in 1919. But Einstein was motivated by the
quest for simplicity and unity. When he announced his
new work, he commented that "scarcely anyone who
has fully understood the theory can escape from its
magic." Herman Weyl, a contemporary and mathematical colleague of Einstein, described it as "the greatest
example of the power of speculative thought"; and Max
Born, one of the fathers of quantum physics, said it was
"the greatest feat of human thinking about Nature."
Had it not been for Einstein, an equally comprehensive
theory of gravity might not have come until decades
later, and been approached by a quite different route.
Einstein is unique among scientists of this century in
the degree to which his work retains its individual
identity.

Indeed, general relativity was put forward so far in
advance of any real application that it remained, for
forty years after its discovery, an austere intellectual
monument, a somewhat sterile topic isolated from the
mainstream of physics and astronomy. This is in glar-
ing contrast to its present status as one of the liveliest
frontiers of fundamental research. Dramatic observational advances that suggest that black holes may exist,
and that made terms such as *quasar, pulsar,* and *Big
Bang* part of the general vocabulary, have brought Einstein's master work in from the cold to the mainstream
of modern research.

Einstein's theory is crucially important to *cosmology*, which is the description of our Universe as a single dynamic entity.* Scientific cosmology is, when you think about it, a rather unusual branch of science. Cosmologists study a unique object—the Universe—and a unique event—the Big Bang. No physicist would be happy to base a theory (let alone a career) on a single, unrepeatable experiment; no biologist would formulate general ideas on animal behaviour after observing just one rat running through a single maze once. But we cannot check our cosmological ideas by applying them to other universes. Nor can we repeat the past evolution of the Universe—although the fact that light travels at a finite speed does allow us to sample the past, by looking at very remote objects, which we see by light that left them long ago. Despite these handicaps, scientific cosmology has proved possible. The reason for its success is that the observed Universe, in its large-scale structure, is simpler than we had any right to expect.

In the 1920s, when cosmologists first devised mathematical descriptions of the Universe ("cosmological models") using Einstein's equations of relativity, they assumed simplicity in order to make the equations tractable. The surprising thing is that those models, deliberately chosen to be as simple as possible, remain relevant today; as observational techniques have improved, they have shown that the Universe itself is as simple as the models—that it is the same in all directions (isotropic) and, as far as we can tell, almost the same everywhere (homogeneous), on the broad picture. (Galaxies are grouped in clusters and superclusters, but even the largest superclusters are still fine-scale detail compared to the entire observable Universe.) It is those equations, combined with observations of the way gal-

*We use the term *cosmogony* for the study of how individual *parts* of the Universe, such as galaxies, came to be as they are.

axies recede from one another, and measurements of
the background radiation that fills all of space, that
lend weight to the idea that the Universe began in a
Big Bang.

The cosmological evidence has strengthened over the
past two decades, but it is still conceivable that our
satisfaction with the Big Bang model will ultimately
prove as illusory and transitory as that of a Ptolemaic
astronomer who, believing the Earth to be at the center
of the universe, successfully fits a new epicycle to the
motion of a planet. Nobody can go back in time to
study the Big Bang itself, but we can learn about it by
studying "fossils" from the earliest eras, just as a geolo-
gist or paleontologist can infer the early history of the
Earth by studying the record in the rocks today. The
theory can never, of course, be "proved," but it is certainly more plausible than any equally detailed alternative model, and we certainly believe it has a better
than even chance of survival.

Since general relativity, and the Big Bang model
built with its aid, provide the best description available of how the Universe got to be the way it is, they
also provide the best basis for investigating how the
Universe will develop in the future. Will it expand
forever, and the galaxies fade and disperse as time
goes by? Or will it one day recollapse, with the sky
falling in to re-create a fireball like the Big Bang?

The answer depends on how much gravitating stuff
there is in the Universe. Imagine a big sphere or asteroid that is shattered by an explosion, the debris flying
off in all directions. Each fragment feels the gravitational pull of all the others, causing deceleration. If the
explosion was sufficiently violent, the debris would fly
apart forever, but, because of this gravitational effect,
at an ever-decreasing rate. However, if the fragments
were not moving quite so fast, gravity would bind them
together, so that eventually the expansion would halt

and the fragments would then fall back together. According to general relativity, much the same argument applies for the Universe.

If the expansion of the Universe is to continue forever, then the curvature of spacetime, described by Einstein's equations, is of a form called, for obvious reasons, "open." If the Universe is destined ultimately to recollapse, then spacetime is said to be "closed." The balance point between these two possibilities, when the expansion goes on forever but eventually *just* comes to a halt (with each part of space ending up as empty and static), corresponds to flat spacetime, or to a "flat universe" model. It is easy to calculate how much matter is needed for its total gravitational influence to bring the universal expansion we observe today to a halt; it works out at about three atoms per cubic metre of the Universe today. Long ago, when the Universe was young and expanding more rapidly, a greater density of matter would have been needed for gravity to exactly balance the faster expansion. But, of course, long ago the Universe *was* more dense. As you might expect, the equations show that provided the Universe starts out with more than the critical density it would recollapse, and at any epoch we could calculate what fraction of the cycle had elapsed if we knew the fractional amount by which the density exceeded the critical value; similarly, if the initial density were too low to stop the expansion, then during the evolution of the Universe it would stay below the critical value. Indeed, as the Universe expands, the actual density always moves further away from the critical density as time passes.*

---

*Such mathematical models of what the Universe *might* be like are usually referred to as "universes," with a small *u*. The capital is, strictly speaking, reserved for discussion of the way our Universe is observed to be. We can talk of open or closed universes, meaning mathematical models, even though we do not know for certain whether *our* Universe is open or closed.

Dark Matter Does the Trick

This is what bring us up against the astonishing cosmic coincidence mentioned earlier. The density of visible matter—bright stars and galaxies—in the Universe today can be inferred by counting the number of galaxies in our region of space, and also by measuring the way in which galaxies move. Changes in the light from distant galaxies (redshifts and blueshifts) tell us how fast the galaxies are moving, both as part of the Universal expansion and through space as they orbit around one another in groups called clusters. Just as the speed of the Earth in its orbit around the Sun is related to the mass of the Sun, so the relative speeds of different galaxies in a cluster tell us how much matter the cluster contains. Putting all this dynamical evidence together, cosmologists find that, in very round terms, there is enough matter in the Universe to provide one-tenth of the "critical" density.

Not all this matter is visible; the dynamics show that the total "luminous" mass of galaxies (bright stars and gas) is barely 1 percent of the critical density, and falls far short of what is needed even to hold big galaxies and clusters of galaxies together. And this may not be the end of the story; there could be more dark matter, in the seemingly empty spaces between clusters of galaxies, which would not show up by these tests but that would contribute to the overall slowing down of the Universal expansion. Just how much more dark stuff there may be is a matter of fierce debate today among the experts. No cosmologist of our acquaintance believes that the Universe contains as much matter as ten times the critical density; most would regard twice the critical density as a wild overestimate.

Studies of the overall expansion of the Universe confirm these estimates, as far as they go. The light reaching us from the most distant observed galaxies has spent several billion years on its journey, and can therefore reveal how fast the Universe was expanding in the remote past. By comparing the recession velocities of more-distant galaxies with those of nearby galaxies, cosmologists could, in principle, infer how quickly the expansion is decelerating and so deduce whether it will eventually halt and reverse. In practice, these measurements show only that the Universe sits so closely on the dividing line—it is so nearly flat—that we cannot tell on which side of the line it lies. Today, 15 billion years after the Big Bang, the density of the Universe is within a factor of ten (between one-tenth and ten times) of the critical value corresponding to a flat universe. And yet, for 15 billion years this density parameter has been steadily moving further away from the critical value! How close must it have been in the beginning, if it is still only a factor of ten away?

The calculation is one of the easiest to carry out using the cosmological equations. It tells us that one second after the moment of creation, the Universe must have been flat to within a factor of 10^{15}—that is, the amount by which the density differed from the critical value, one way or the other, was by a decimal point followed by 15 zeroes and a 1. How much further back in time the laws of physics as we know them can be applied is, to some extent, uncertain. But in quantum physics there is a fundamental limit to the accuracy with which time can be described—in a sense, the "quantum" of time. This unit, the Planck time, is 10^{-43} of a second. Cosmologists today, drawing on theories developed by particle physicists, are attempting to describe the actual origin of the Universe in terms of quantum events that happen on this sort of time scale. Such theories, which we discuss in more detail later, are as yet far less well established than the standard model of the Big Bang; nevertheless, if we take them at face value we deduce that the Universe has been expanding since 10^{-43} seconds after "time zero"—but we have no way of knowing what went on between time zero and 10^{-43} seconds. Pushing back to that moment, the nearest definition we have to a beginning, the flatness of the Universe must have been precise to within 1 part in 10^{60}. This makes the flatness parameter the most accurately determined number in all of physics, and suggests a fine-tuning of the Universe, to set up conditions suitable for the emergence of stars, galaxies, and life, of exquisite precision.

If this were indeed a coincidence, then it would be a fluke so extraordinary as to make all other cosmic coincidences pale into insignificance. It seems much more reasonable to suppose that there is something in the laws of physics that requires the Universe to be *precisely* flat. After all, the critical density for flatness is the *only* special density; no other value has any cosmic significance at all. It makes more sense to accept that the Universe had to be born with *exactly* the critical expansion rate than to believe that by blind luck it happened to start out within 1 part in 10^{60} of the critical value. Physicists apply a similar argument when they say that the mass of a photon, a quantum of radiation, is precisely zero. No experiment can measure a precisely zero mass; the best that can be done is to set a limit, from experiments, and say that the mass must be less than 10^{-58} of a gram. In both cases, we infer that there is *no* deviation from the interesting value of the number.

There is, indeed, a theory (rather, a group of theories) that makes flatness a required feature of the Universe. Such models go by the name *inflation*, because they suggest that very early in its lifetime (during the first split-second) the Universe expanded by an enormous factor, with a volume of spacetime smaller than a proton expanding to the size of a basketball in a tiny fraction (about 10^{-35}) of a second. This phase of rapid accelerating expansion, or inflation, would have smoothed out any wrinkles in spacetime and flattened the fabric of the Universe. The process is the opposite of what happens to a smooth plum when it dries up to become a wrinkly prune; any wrinkles in spacetime got smoothed away during inflation, according to these models.

Inflation is interesting in its own right, and the exotic physical processes that occurred at ultra-early stages have implications that we will discuss later. From our immediate point of view, however, it is important because it offers the best physical reason *why* our Universe should be *exactly* flat. The implications of this are profound indeed, and at one level seem to remove humankind further than ever from the centre of the cosmic stage.

As we have mentioned, studies of the way galaxies move within clusters show that clusters contain about one-tenth of the matter required to make the Universe flat. These estimates correspond to about 0.3 atom per cubic metre, and are in fact quite nicely in line with calculations of conditions during the Big Bang (starting at about one-hundredth of a second after the moment of creation and going up to the end of the first four minutes) required to produce the right mix of hydrogen, helium, and deuterium. Those calculations show that the amount of energy that could have been processed into protons and neutrons in the Big Bang was only about one-tenth of the amount needed to make the Universe flat, perhaps a little less. For twenty years or so, most cosmologists took this at face value, to mean that the Universe must be open. They simply never considered seriously the possibility that there might be other forms of matter besides protons and neutrons (and their associated, but lightweight, electrons) in the Universe.

In the 1980s, however, some theorists began to worry

much more about the cosmic coincidence implied by the near-flatness of the Universe today, while particle physicists began to suspect that there might be other forms of matter allowed (even *required*) by the laws of physics, which could have been created in large quantities during the Big Bang. Just because our own bodies, planet Earth, the Sun, and all the stars in the sky are composed of protons and neutrons (collectively dubbed *baryons*), cosmologists realised, that was no proof that *all* of the stuff of the Universe had to be in the form of baryons. The argument that we are made of baryons and therefore the Universe must be made of baryons is in fact as anthropocentric and unfounded as the argument that because we see stars surrounding the Earth on the bowl of the sky, the Earth must be the centre of the Universe. Indeed, if the Universe were made only of baryons, in the numbers that theory tells us were produced in the Big Bang, and contained no nonbaryonic dark stuff at all, then matter would be spread so thin that galaxies (and clusters of galaxies) almost certainly could not have formed in the way we see them around us. Without the dark stuff, galaxies, and ourselves, might not exist at all.

So, the answer to the question posed by the title of this chapter must be that the Universe *is* a special place, balanced on a knife edge between being open and closed. We cannot be sure what made it special—why the laws of physics require it to be flat—but this line of argument tells us that there is more to the Universe than the kind of atoms that make up our own bodies and the protons and neutrons that form the stuff of stars. At least 90 percent of the stuff of the Universe is in the form of dark matter, which cannot all be made of baryons. And yet, without the dark matter—the invisible stuff of the Universe—the Universe itself would be a very different place, and we would not exist. Does the dark stuff exist, in a sense, for our benefit? Is it there because we are here? What is it? Where is it? And how has it been responsible for the emergence of the kinds of structures we see in the Universe?

CHAPTER TWO

---★---

The Geography of the Universe

AS WE HAVE SEEN, astronomers and cosmologists deal with billions of light-years of space, and billions of years of time. Our own Sun is 4.5 billion years old, and is destined to evolve for at least as long again before its nuclear fuel runs out. The geography of the Universe is the geography of both space and time, because when we look out into space we see things as they were when the light now reaching our telescopes actually left the objects we are studying. Even the Sun, our nearest star, is so remote that its light takes more than eight minutes to reach the Earth.

Clearly, astronomers cannot learn about the life cycles of stars by watching an individual star, like our Sun, live out its life. But just as a botanist could find out about the life cycle of a tree by wandering about a forest and examining trees in different stages of growth, so an astronomer can infer the life cycle of a star by studying many stars of different ages. Stars like the Sun start their lives by condensing, under the inward tug of gravity, from interstellar clouds. After a few hiccups, signs of youthful exuberance, they settle down into a state where they are kept hot by the steady fusion of hydrogen into helium in their interiors. During this phase of quiet life, which our Sun is now about halfway through, a star is said to be on the "main sequence."

When hydrogen in the core of a star like our Sun is exhausted, however, it first swells up to become a red giant, then shrinks in upon itself, cooling into a ball roughly the size of the Earth, called a white dwarf.

All this is well understood. But not everything in the cosmos happens on quite such a leisurely time scale, and with so little excitement. Some stars, which are much heavier than our Sun, end their lives by exploding violently, as supernovae. Such events are relatively rare—in 1987, astronomers were excited by the opportunity to study the first "nearby" supernova seen since the invention of the astronomical telescope in the seventeenth century; but even this "nearby" event was 170,000 light-years away, in a neighbouring galaxy to the Milky Way, known as the Large Magellanic Cloud. The light that astronomers saw in the supernova in 1987 had set out on its journey before the beginning of the most recent ice age on Earth. Nevertheless, supernovae are important. They mark the violent end point of stellar evolution, when a star too massive to become a white dwarf exhausts its available nuclear energy. The core of the star collapses catastrophically (it "implodes"), while the outer layers are blown away into space; what is left is a dense stellar cinder, a *neutron* star only 10 kilometres across, but containing about as much matter as our Sun.

In such a star, material is squeezed to nuclear densities, 10^{14} times greater than the density of ordinary solids. The gravitational force on the surface of a neutron star is 10^{12} (a thousand billion) times greater than on the surface of the Earth, and in order to escape from the gravitational pull of such a star a rocket leaving its surface would have to be fired upward at about half the

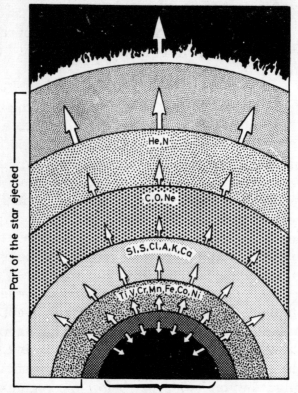

Figure 2.1 The "onion skin" structure of a massive star before it explodes as a supernova. The hotter inner shells have been processed further up the periodic table; this releases progressively more energy until the material is converted into iron, the most tightly bound nucleus. Endothermic nuclear reactions occurring behind the shock wave that blows off the star's outer layers can synthesise small quantities of still heavier nuclei.

speed of light. The conditions inside a neutron star are more extreme than those in the stars we see shining in the sky. In a neutron star, gravity has squeezed out of the system all memory of its original nuclear composition. The internal structure is still poorly understood, because the conditions are so exotic and unfamiliar. But neutron stars provide a good example of the way in which physicists can test their theories—perhaps to the

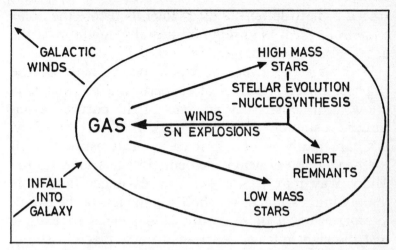

Figure 2.2 Processes whereby the content of a galaxy gets gradually converted into heavy elements and long-lived stars.

breaking point—by applying them to the behaviour of matter under extreme conditions. The Big Bang was a time when conditions were even more extreme; and while the relevant physics is still uncertain, it may prove to be simpler than that of neutron stars—there were fewer options available for matter then.

Supernovae, even the nearest ones, may seem a long way away and a long time ago. But it is only by studying such events that astronomers can tackle such an everyday question as *where the atoms we are made of came from*. Complex chemical elements are built up from hydrogen, as we shall explain in chapter 10, by the nuclear reactions that provide the power source in the cores of ordinary stars. When solar-type stars die, the elements that have been built up stay there, as part of the cooling white dwarf. But when massive stars explode as supernovae, they scatter traces of heavy elements into their surroundings. Some of those atoms have become us. Biologists can trace our ancestry back to primitive protozoa, but the astronomer goes back further still. Each carbon atom in your body can be traced back to stars that died violently before the Solar System formed. We are, quite literally, made from the ashes of long-dead stars.

Without supernovae, we would not be here. So the rules of physics that allow supernovae to exist also allow us to exist; they are part of the pattern of anthropic cosmological coincidences. All of this, however, is a parochial view of cosmic events. Interesting though the production of heavy elements in supernovae in the Milky Way may be for our own existence (and it is doubly interesting, as we shall see, in view of the exquisite precision of those cosmic coincidences that allow just the right nuclear reactions to take place in stars), our whole Milky Way Galaxy is tiny in the perspective of the whole known Universe. A galaxy like our own may contain a hundred billion (10^{11}) stars, and millions of those stars may, for all we know, have retinues of life-bearing planets. But to a cosmologist a galaxy is simply a speck in space, regarded as a "test particle" that is useful primarily as a marker by which the cosmologist can measure the rate at which space is expanding. The redshift is, indeed, the key to measurements across the Universe; it is the cosmologist's equivalent of the geographer's theodolite.

Redshifts, Galaxies, and Quasars

Observational cosmology became possible in the 1920s, when Edwin Hubble discovered a key to the Universe. He found that for all but the closest neighbours of the Milky Way (such as the Large Magellanic Cloud, which is held in orbit around our Galaxy by gravity), the redshift of a galaxy is proportional to its distance from us. This discovery was part of a revolution in understanding the cosmos in which, for the first time, scientists realised fully that our Milky Way is just one

ordinary, fairly typical galaxy similar to millions of
others, and that galaxies are the basic units making up
the large-scale Universe. They are systems of stars held
in equilibrium by a balance between the effect of gravity, which tends to make the stars fall together, and the
countering influence of stellar motions, which, if gravity did not act, would make the system fly apart. In
some galaxies, like our own, the stars move in nearly
circular orbits in giant discs; in others, the less photo-
genic *elliptical* galaxies, the stars swarm about in more
random fashion, each feeling the gravitational pull of
all the others.

Galaxies are to astronomers what ecosystems are to
biologists. They are not only *dynamical* units, held together by gravitation, but act as *chemical* units as well.
The atoms we are made of come from all over our
Milky Way Galaxy. They were forged in many different
stars, and they may have spent a billion years or more
wandering in interstellar space before finding themselves in the gas cloud that became our Solar System.
But few of the atoms in our bodies come from other
galaxies. Each galaxy experiences its own ongoing evolution as new stars—the "organisms" of the galactic
ecosystem—continue to form from the debris of their
predecessors.

The light from most galaxies is due essentially to the
stars and gas they contain. Stars in other galaxies are
too faint to be detected individually (except in our
nearest galactic neighbours), but the light from billions
of stars combines to make a fuzzy patch in the field of
view of a telescope, converted into a bright image only
by long exposure on an astronomical photograph. That
fuzzy patch of light can be studied spectroscopically, and
the positions of familiar features in its spectrum—the
distinctive pattern of colours from sodium, for instance—
can be determined. Even without knowing *why* the
redshift should be proportional to a galaxy's distance

from us (a fact now known as Hubble's law), cosmologists could make use of the discovery. The "law" was first established by measurements of light from relatively nearby galaxies, whose distances can be estimated by other means. For most of the fuzzy blobs they photograph, astronomers have no direct means of estimating distance. But Hubble's law is corroborated by the evidence that galaxies of the same standard type, which are presumed to have the same intrinsic brightness, appear fainter when their redshifts are larger—as they should if redshift is a measure of distance. Insofar as the law is correct, then *distance is proportional to redshift*. Turning Hubble's law around, astronomers are now able to estimate the distance to any galaxy whose spectrum they can take, simply by measuring its redshift.

There are still uncertainties in all this. The distances to nearby galaxies are only estimated by a complex chain of arguments. Further uncertainty comes in because galaxies don't move exactly with the Hubble "flow" but also have their own "peculiar velocities" of a few hundred kilometres per second. These velocities are a substantial part of the overall velocity for the relatively nearby galaxies whose distances are independently measurable, which are the very ones used to relate redshift and distance. This relationship is consequently still uncertain by a factor of two; so some cosmologists would estimate all distances across the Universe to be twice as big as the estimates favoured by other cosmologists. Hubble's law also implies a *timescale*—the time it would have taken for galaxies to move apart to their present positions, each at the constant speed indicated by their redshifts, if they started out touching one another. This time is between 10 and 20 billion years. To alleviate the uncertainty, most experts include in their equations a factor dubbed *h*, which is 1 if the "Hubble time" is 10 billion years and 0.5 if it is 20 billion years; that way, the numbers the cosmologists give you can be instantly adjusted, by adjusting h, to match one's favoured value for the cosmic distance scale.

Another problem is understanding *why* the redshift-distance relation should hold. The standard explanation is the one we have already mentioned in passing. Light from an object that is moving fast enough away from you will be redshifted, by a process known as the Doppler effect; indeed, the same thing happens to sound waves in the air when the source of the sound waves speeds away, deepening the note of a train whistle or the siren of a police car. Similarly, an object moving towards you will produce blueshifted light, with shorter wavelengths, or a higher pitch from the same siren that sounds deeper when the car is moving away. The waves are more spaced out, or more crowded together, simply because of the motion of the object that emits them. This Doppler shift (both red and blue) is very useful throughout astronomy; it can tell us something about how stars and gas clouds are moving within our Galaxy, how other galaxies rotate, and how galaxies in a cluster move relative to one another.

There is another, equivalent way to envisage Hubble's law. The Universe (that is, the space between the galaxies) is expanding, and can be envisaged as carrying galaxies along with it. As light travels through expanding space, it is stretched to longer wavelengths, and longer wavelengths are those at the red end of the spectrum. In a uniformly expanding universe, the effect will be bigger for more distant galaxies, and redshift will indeed be proportional to distance.

A different way of producing a redshift is with the aid of a strong gravitational field. Light struggling out from the surface of a neutron star, for example, has to work so hard against gravity that it is redshifted. Light cannot be slowed down, but it can lose energy. As well as having a longer wavelength than blue light, red light has less energy.

These processes are well understood. But a few people worry that there may be still other ways to produce redshifts, which we do not yet understand, and that some of our ideas about the geography of the Universe may be at fault because we have placed too much faith in Hubble's law. These concerns usually centre on the objects that include, if the Hubble's law interpretation of their redshifts is correct, the most distant and most energetic entities ever observed—the quasars.

Galaxies show up as fuzzy patches on astronomical photographs, quite different from stars, which look like pinpoints of light. But in the 1960s astronomers discovered that some of the bright pinpoints of light that they see in their telescopes and photograph with astronomical cameras have very large redshifts, comparable to those of distant galaxies. Because they look like stars but have redshifts appropriate for galaxies, these sources were called *quasistellar objects*, which was soon contracted to *quasar*. Quasars look like stars because they are very small. But if they are at the distances implied by the Hubble law interpretation of their redshifts, they must also be very bright—as bright as a galaxy that contains a thousand billion stars, in some cases far brighter. All of the energy that makes a quasar shine so brightly must be coming from a very small region, a volume of space no bigger than the distance across our Solar System. Although a quasar produces more energy than our whole Milky Way Galaxy, it would fit within the orbit of Pluto around the Sun.

Some astronomers, in the 1960s, found this too much to swallow. They suggested, instead, that the redshifts seen in quasars are not due to the stretching of space-time as the Universe expands, but are familiar Doppler shifts. On that picture, it was argued, quasars might simply be starlike objects (not *quasi*stellar at all) that had been shot out from the middle of a nearby galaxy at very high speeds. The argument fell down for several reasons. For example, more and more quasars were discovered, but none showed a blueshift, although surely some fragments of a nearby cosmic catastrophe would have been moving towards us. And as observing instruments have improved, astronomers have now obtained evidence that many quasars, perhaps all, are actually embedded within galaxies; they have come to realise that there is a whole range of violent activity that occurs in the centres of galaxies, with quasars just the most extreme example. In a handful of cases, beautifully precise measurements have been able to take the redshift of a quasar in the nucleus of a galaxy and, quite separately, to measure the redshift of the material of the galaxy outside the nucleus. Both redshifts are the same, and the spectrum of the light from the faint outer regions resembles that of light from the stars and gas of an ordinary galaxy. There is other evidence that supports the Hubble law for quasars, as well. On balance, the general interpretation of quasar redshifts as an effect of the expanding Universe—as cosmological redshifts—seems sound. But we should, in fairness, mention the dissenting minority who believe that the conventional "cosmological" view of quasars is as distorted and incomplete as Ptolemy's picture of the Solar System.

Most investigations into the nature of the Universe, including the main themes of this book, deal with the broad brush strokes of the cosmic picture, the general view of what is going on in the cosmos. A few cosmologists, however, prefer to focus their attention on peculiarities and oddities that do not easily fit in to the broader picture. This is always the way in science. Sometimes, studies of the oddities turn out to reveal new insights that lead to a redrawing of the broad picture; more often, the oddities become incorporated into the overall scheme as a fuller understanding develops. The particular redshift oddities that worry a few researchers deeply concern the appearance on the sky (and therefore in astronomical photographs) of objects that have different redshifts but seem to be physically connected to one another. The best examples of these peculiar connections have been photographed by Halton Arp, an American astronomer who is now based in Germany. He has obtained pictures of systems where these "discrepant" redshifts occur, in chains of galaxies and in situations where a quasar seems to lie at the end of a jet of material shot out from a galaxy.

At one extreme, a few astronomers argue that these pictures cast doubt on the whole idea of cosmological redshifts. Obviously, if two objects are physically connected then they are at the same distance from us, and should, if Hubble's law is a guide, have the same redshift. If even one redshift is "wrong," runs the argument, perhaps they all are. At the other extreme, some astronomers dismiss all of Arp's photographs as coincidences. The "bridge" seemingly joining a galaxy to a quasar with a different redshift is, they argue, always an optical illusion, and there is no problem at all. That dismissal was easy when Arp had found only one or two peculiar associations, but becomes less tenable with every new "coincidence" he turns up.

We stand somewhere in between the two extremes. Along with most astronomers who have followed Arp's work, we judge that the case for anomalous redshifts has not gained strength over the years, and has even weakened as extragalactic astronomy has advanced. Arp himself attributes this scepticism to a blinkered antipathy on the part of his colleagues towards radical new ideas. A *few* astronomers, especially some who have staked years of effort on research programmes based on "conventional" cosmological assumptions, might indeed be psychologically indisposed to accept that the standard picture might be wrong. But *most* would surely be *delighted* at the prospect of uncovering fundamentally new phenomena and even "new physics." Astronomers are generally only too eager to embrace novel ideas in whose further exploration they can share. The reluctance of astronomers to devote their observing time to following Arp's lead is perhaps analogous to the reluctance of most scientists to study ESP; if such phenomena were real, there would be a colossal payoff, but the probability seems so minuscule that the most open-minded enquirer is wary of investing effort. Arp has, however, now gathered so much evidence of peculiar associations that it is hard to deny that, in some cases at least, there is more than coincidence and optical illusion at work. Something odd *is* going on, in some cases, but it by no means follows that conventional ideas will not be able to explain these phenomena when they are better understood. In astronomy, as in other sciences, phenomena often elude our understanding for decades, even when they eventually turn out to be fully explainable in terms of known laws. So it seems premature to throw in the sponge and invoke new physics until astronomers have thought longer and harder about these associations.

However the details may change, the broad picture seems secure—there are now thousands of quasars known, and relatively few show the peculiarities Arp is fascinated by. Dimmer quasars, by and large, have larger redshifts, which is what you would expect if large redshift implies large distance, and there is a nice gradation of activity seen in galaxies, from quiet ones like our own through various kinds of energetic outbursts right up to quasars. It is disappointing, in a way, that we are not in the midst of a revolution in our understanding of redshifts, a revolution that would be as exciting for astronomers as the revolutionary discovery by Hubble that the Universe is expanding. But we

shouldn't be greedy—after all, we *are* living through
another revolution in our understanding of the Universe, which is the main theme of this book.

All of the debate about anomalous redshifts is, to
some extent, a side issue in our present discussion. The
geography of the Universe as we know it is primarily
the geography of the distribution of galaxies. Even Arp
does not suggest that there is any real doubt about the
redshift-distance relationship as applied to most of the
galaxies studied by astronomers; so whatever is happening in a few peculiar objects, we can still map out
the visible region of the Universe with some confidence
by measuring redshifts, and therefore distances, to large
numbers of galaxies. When we do so, we find surprises
as exciting and far-reaching as any of the implications
of Arp's interpretations of peculiar redshifts. We find
confirmation that 90 percent of the Universe is not in
the form of bright stars and galaxies, and clues as to
what form it is in. This unseen dark matter dominates
the geography of the Universe and has set the scene for
the emergence of intelligent life on at least one planet.
We shall discuss its nature shortly. But first, for completeness, we should make the best use we can of measurements of quasar redshifts, since the largest of these
are far larger than any measured redshift for a normal
galaxy, and provide the cosmic geographer with a distant view of the edge of our Universe.

To the Edge of the Universe

Redshift is, by everyday standards, a rather peculiar
measure of distance. By convention, astronomers denote the redshift of an object by the letter z (which
denotes the fractional increase in the measured wavelength of light). When they map out the distances to

Figure 2.3 Escher's infinite lattice.

galaxies in various directions around the Milky Way Galaxy, the values of z they measure are usually small-ish fractions—a redshift of 1 is big for a galaxy, although astronomers using new techniques in the 1990s should readily be able to identify galaxies even farther away and measure their redshifts. Hubble's law, that redshift is proportional to distance, is based on measurements of nearby galaxies. If all the rods in Escher's infinite lattice (figure 2.3) were to lengthen at the same rate, the lattice would keep its shape but would expand—space would stretch. But there is no centre. An observer at *any* vertex would see all other vertices receding, the recession rate being faster for the more distant ones, in accordance with Hubble's law. This is a good analogy to the expanding Universe—except, of course, that the galaxies are actually scattered in a complicated pattern of groups and clusters, rather than being equally spaced.

General relativity confirms that this is the redshift law that "ought" to be seen for nearby galaxies in a universe expanding in line with Einstein's equations. But general relativity also tells us that this is only an approximation to a more complicated rule that applies in general across the Universe. For fairly nearby galaxies, redshift is indeed closely proportional to distance. The farther out we look into space, however, the more the redshift law deviates from this simplicity. Using general relativity, we can still calculate distances to remote galaxies, and quasars, by measuring redshifts. But a redshift of 2 does not mean that a galaxy is exactly twice as far away from us as a galaxy that has a redshift of 1. Astronomers are wary of quoting distances to galaxies in terms of light-years because of the uncertainty in their estimates of the constant that appears in Hubble's law, and in the relativistic version of the law. It is better to envisage the redshift as a measure of how much the Universe has expanded—how much the wavelengths have stretched—since the epoch when the light set out on its journey towards us.

If a galaxy or quasar has redshift z, the scale of the Universe (the separation of two typical galaxies) is now $(1 + z)$ times as large as when the light set out. So if $z = 3$, for example, the expansion factor is 4. How does the redshift relate to the look-back time—the time since the light set out? If the galaxies had always moved at the same speed, the answer would be straightforward: When the Universe was a quarter of its present scale ($z = 3$), it would have been one-quarter of its present age, and we would be looking back three-quarters of the way to the Big Bang. More generally, a redshift z would correspond to $1/(1 + z)$ of the present age. Moreover, the age of the Universe would simply be the Hubble time, $10/h$ billion years. However, there is no reason to expect that the galaxies have always moved at their present speeds; indeed, we expect a *deceleration*, due to the gravitational pull of each galaxy on every other. The average speed of galaxies over their past history must have been higher than the speeds we measure today using their redshifts. This somewhat reduces our estimate of the time since the Big Bang.

The deceleration also changes the relationship between redshift and look-back time. For our favoured flat Universe, the relation is fairly simple: The age of that universe at a redshift z is not $1/(1 + z)$, but about the three-halves power of this fraction (the square root of the cube of the number). So when we look at a quasar with a redshift of 3, we see it as it was when the Universe was one-eighth of its present age ($4^{3/2} = 8$) — when it was less than 2 billion years old (figure 2.4).

This simple calculation highlights a fundamental cosmic puzzle: How could the Universe have been so smooth and uniform long ago? When the Universe was one-quarter of its present size, it was only one-eighth of its present age. So there had been proportionately less time available for any influence, which can never travel faster than light, to spread across the Universe. Regions that cannot "communicate" with each other cannot come into synchrony, so why are different parts of the Universe so similar to one another? The closer we probe back towards the Big Bang, the worse this problem becomes— as a fraction of the time light takes to cross the Universe, there is less and less time available for the different parts of the Universe to interact with one another. This causality problem, highlighted in figure 2.4, arises because gravity *decelerates* the cosmic expansion. The inflation hypothesis, which we shall discuss later, postulates an early stage of rapidly *accelerating* expansion, and offers an explanation of why the Universe was very uniform even in the earliest stages of the Big Bang.

As we look farther out across the Universe we are also, of course, looking back in time, seeing the Universe as it was when the light we see left those galaxies.

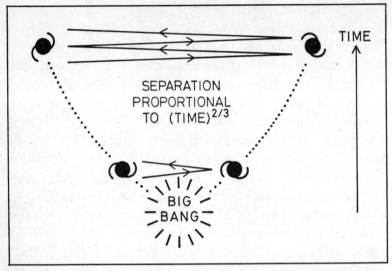

Figure 2.4 Communication between different parts of the Universe was *more difficult* at earlier times. Consider, for instance, a galaxy whose distance from us is ¼ of the Hubble radius. We can now exchange light signals 4 times during the expansion timescale of the Universe. However, when the Universe was ¼ its present size (redshift $z = 3$) it was ⅛ its present age; therefore, even though this galaxy was then 4 times closer, there would have been time for only 2 signals to be exchanged. Extension of this line of argument implies that no two galaxies (or protogalaxies) would have been in causal contact at very early times. It is therefore a mystery why the Universe started its expansion in such a uniform and apparently well-synchronised fashion. In this example, the time since the Big Bang is actually only ⅔ of the Hubble time, because expansion was faster in the past.

But the redshift imposes a kind of barrier between us and the Big Bang itself. Each doubling of the measured redshift does not double the measured distance, either in terms of space or of time. Instead, the farther away you look—the farther back in time—the bigger the step in redshift you need to cover a chosen distance. An imperfect, but helpful, analogy is with a climber who tackles a high mountain. At first it is easy to make progress, and for a small effort the climber gains a lot of height. As the mountaineer gets higher and the route gets steeper, however, that same amount of effort brings

Figure 2.5 Because of the finite speed of light, we observe remote regions as they were at early epochs when everything was closely packed together. The Universe as we actually see it resembles this Escher picture: Objects seem more and more crowded together towards our observational "horizon."

ever-diminishing returns. In the case of the Universe, the Big Bang itself, the moment of creation, is at *infinite* redshift, and can never be seen directly. It is as if you can always get a little bit farther up the mountain, if you try hard enough; but to get to the very top requires infinite effort.

Quasars have now been found with redshifts of about 4.5, and in round terms this means that we are looking back across more than 90 percent of the history of the Universe, to a time only about a billion years after the Big Bang. Increasing the redshift record to $z = 10$ would push us back to when the Universe was 3 percent of its present age. We don't know whether galaxies had formed by then; because of the time required for galaxies to form after the Big Bang, we may already be close to "seeing" the practicable "edge of the Universe," an "edge" that has the rather peculiar property that it corresponds to a time when the Universe was smaller, and more densely packed, than it is in our neighbourhood today.

This "edge" doesn't imply in any sense that we are in the middle of the Universe. A better analogy is with a sailor who seems, from the deck of a ship, to be surrounded by a circle of ocean with a distinct edge, the horizon; if the sailor climbs the mast of the ship, the extra height will give a view of a bigger circle of ocean. The new horizon still looks like an "edge," even if the ocean is boundless.

Cosmologists, in fact, turn the redshift relation around and use it, for high values of z, as a convenient means of labelling the sequence of events in their mathematical models (based on general relativity) of how a universe like ours may have evolved out of the Big Bang. They talk of events that happen at a redshift of 1,000 or 100, or whatever the figure might be, instead of speaking of the time in years after the moment of creation. This is a convenient shorthand, but it does not mean that any astronomer has ever measured a redshift that big, or that there is any hope that such a cosmological redshift could ever be observed—except for the special case of the background radiation that fills the whole of the Universe. This is interpreted as the light of the white-hot Universe, a few hundred thousand years after the Big Bang itself, redshifted so much that it now shows up as a feeble hiss of energy in the microwave part of

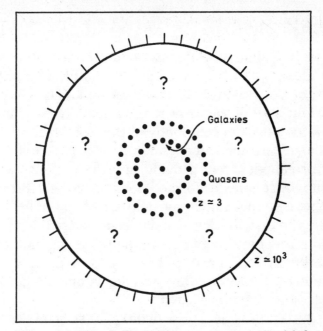

Figure 2.6 This diagram illustrates various "redshift shells" around us. The microwave background comes from the "cosmic photosphere" at $z = 1,000$, corresponding to an epoch when the Universe was about a million years old. The most distant quasars emitted the light now reaching us when the Universe was about a billion (10^9) years old. We know very little about the era of cosmic history between 10^6 and 10^9 years.

the electromagnetic spectrum. This corresponds to a redshift of about 1,000. Nothing has been detected from the part of cosmic history corresponding to the gulf between $z = 1,000$ and $z = 5$, covering about 6 percent of the history of the Universe. Somewhere in that region, the processes that led to the formation of galaxies took place. Lacking any direct observations of those processes at work, we have to work out how galaxies formed by looking at the way they are distributed in the Universe today. Which brings us back from the edge of the Universe to the geography of our own cosmic neighbourhood.

The Bright Stuff

We live in a galaxy. Galaxies contain stars, and stars are made of baryons—the same sort of stuff, to a physicist, as our own bodies are made of. Galaxies, the visible, bright stuff of the Universe, are also the basic units we use to study the geography of the Universe, but not all galaxies are the same. Differences between various kinds of galaxies may provide important clues to the way in which galaxies formed, long ago when the Universe was young, and may help us to estimate how reliable the distribution of galaxies is as a guide to the distribution of all the matter in the Universe, including the dark matter. If galaxies had not formed, we would not be here. Understanding galaxies is one of the keys to understanding our existence.

Our Milky Way is a disc galaxy. Disc galaxies are also known as spirals, because the pattern of bright stars in such a galaxy often forms a prominent spiral*; however, not all disc galaxies show pronounced "spiral arms," so we prefer the term *disc*. In addition to the disc itself, such a galaxy has two other distinct components—a central bulge of stars around the nucleus, giving the galaxy something of the appearance of a fried egg, and a halo of old stars surrounding both the disc and the bulge in a huge, roughly spherical distribution. Some of the stars in the halo are grouped together in spherical, or globular, clusters, aggregations of stars that move together through space as a unit. A globular cluster may contain up to a million stars held together by gravity within a radius of 100 light-years;

---

*The pattern is produced by waves moving through the galaxy; each star follows a nearly circular orbit around the centre of the galaxy, and does not move along a spiral "arm." This is rather like the way a pattern of waves moves across the surface of the sea, even though each molecule of water is bobbing up and down as the wave passes, not moving forward with it.

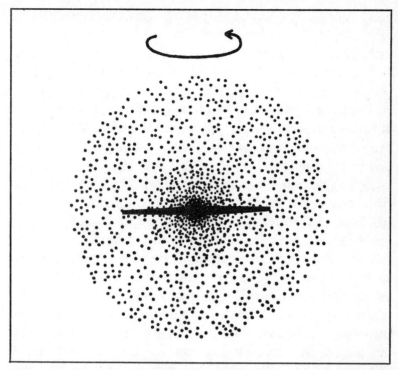

Figure 2.7 Schematic "edge on" view of a disc galaxy like our own Milky Way, showing its three main components: central bulge, disc, and halo.

but there are only about 200 globular clusters in the Milky Way Galaxy. Halo stars are a small fraction of the hundred billion stars that make up a disc galaxy.

Measurements of the way disc galaxies rotate, using the ubiquitous redshift (the straightforward Doppler version, this time) also reveal that there must be a great deal of dark matter in the halo of such a galaxy, holding the bright matter of the disc in place and stabilising the rotation (see chapter 5).

The disc of such a galaxy really is thin, compared with its diameter. In the case of our own Galaxy, for example, it forms the band of light across the sky that gives the Milky Way its name, and it is less than 1,000 light-years thick in the region of our Solar System, which is about 28,000 light-years out from the centre of the Galaxy, two-thirds of the way to the edge of the bright disc. These are fairly typical measurements. Old stars, called Population II, occur mainly in the halo and the nuclear bulge, which extends out about half-way to the position of the Solar System. Younger stars, which have formed out of clouds laced with the debris of former supernovae, lie mainly in the disc. Most of the bright stars in a galaxy like our own are young stars, called Population I.

By measuring the speeds with which other galaxies close to our own, in what is called the Local Group of galaxies, are moving, astronomers can calculate how much mass our Galaxy must have in order for its gravitational influence to explain those movements. (Some of these galaxies are moving away from us; others, such as the Andromeda Galaxy, are moving towards us.) The *minimum* amount of mass in our Galaxy, estimated in this way, is about a thousand billion times the mass of the Sun. Since, in round terms, the Galaxy contains a hundred billion stars and the average mass of a star is close to the mass of our Sun, that means that there is at least ten times more dark matter in the Galaxy than there is in the form of bright stars. This figure is typical of estimates for other disc galaxies—by and large, such galaxies contain ten times more dark matter than bright.

The other main type of galaxy is the *elliptical*. These come in a variety of shapes and sizes. Some are spherical, like an enormous globular cluster; some are elongated, like a football or a cigar; many are oblate, like a slightly squashed sphere. They are all made up almost entirely of old stars, and in many ways they look like the nucleus and halo of a disc galaxy, without the disc. Some "dwarf" elliptical galaxies are very small compared to our Galaxy. At the other extreme, the so-called "CDs" are the largest galaxies known, with their stars stretching out across more than 300,000 light-years from the centre. Such a galaxy may contain up to a hundred

times as many stars as our own Galaxy. Dwarf ellipticals, however, may be the most common galaxies in the
Universe; or that honour may go to another kind of
small galaxy, called dwarf irregulars since they come
in all kinds of irregular shapes.

Getting ahead of our story slightly, it seems that
galaxies must have formed from clouds of gas, chiefly
hydrogen and helium produced in the Big Bang, which
were held together both by their own gravity and by
the gravity of clouds of dark matter in which they were
embedded. The dark matter produced a kind of gravitational pothole, or potential well, into which the gas
settled and became dense enough to collapse under its
own gravity, and condense into stars. Without the gravitational influence of the dark matter, a galaxy like the
Milky Way, with its stars and star systems, might never
have formed. The key question, which we shall address
shortly, is how the initial irregularities, the gravitational potholes, formed in a universe expanding smoothly
away from the Big Bang.

The distribution of galaxies in the Universe today is
by no means perfectly smooth. More than half of all
known galaxies occur in groups. Small groups, which
might contain ten or twenty galaxies held together by
gravity, are called just that—"groups." Larger groups,
which contain hundreds or thousands of galaxies that
lie in the same part of the Universe, are called clusters.
A group of galaxies within a cluster is rather like a
single island in an archipelago. Some clusters are regular in shape, spread over spherical volumes of space
with more galaxies in the centre of the cluster and
relatively few farther out; others have a more lumpy
appearance, and usually these irregular clusters contain a higher proportion of disc galaxies.

Clusters themselves congregate together in superclusters. For example, the Local Group of galaxies, of which
our Milky Way is a member, is part of the Local Super-

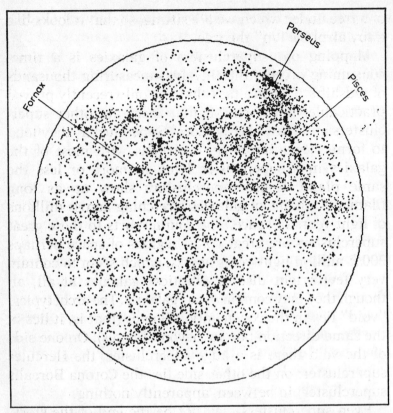

Figure 2.8 This diagram (courtesy of Ofer Lahav) shows the more luminous galaxies in the northern sky out to a distance of 50–100 megaparsecs. Three well-known clusters are marked; the complex and "filamentary" character of the galaxy distribution is evident from this picture.

cluster, a straggling collection of thousands of galaxies that seems to be centred on the Virgo cluster. The Virgo cluster is named after the constellation in which it is seen in the sky—a common astronomical practice. But this group of galaxies is, the redshift test shows us, actually far beyond the constellation Virgo (which is simply a pattern formed by stars within our Milky Way Galaxy), at a distance of about 50/h million light-years from our Galaxy. Its appearance "in" Virgo is a chance alignment of two objects on the sky, rather as we might see a high-flying aircraft passing behind the branches of a tree under which we are sitting, so that it looks like a toy airplane "in" the tree.

Mapping out superclusters of galaxies is a time-consuming process that involves measuring thousands of redshifts. This is a task that has only recently proved practicable; it has shown that there are other super-clusters and that, like the Local Supercluster, they tend to form sheets across the Universe, with most of the galaxies in a supercluster lying in more or less the same plane. Some superclusters show up as long filaments, chains of galaxies stretching across millions of light-years of space. The counterparts to these great superclusters are great voids, regions of space perhaps $200/h$ million light-years across, which seem to contain very few bright disc or elliptical galaxies at all, al-though they may contain dwarf galaxies. The archetypical "void" lies in the constellation Boötes (that is, it lies in the same direction but far beyond Boötes). On one side of the void there is a huge supercluster, the Hercules supercluster; on the other side lies the Corona Borealis supercluster; in between, apparently nothing.

Even superclusters may not be the end of the story. Brent Tully, of the University of Hawaii, was one of the astronomers who first mapped out the extent of the Local Supercluster. This is roughly 100 million light-years across, if $h = 0.5$. In 1987, Tully presented evidence that the whole supercluster may be physically associated with other superclusters, forming a structure he calls the "Pisces-Cetus complex," stretching for more than a billion light-years across the Universe. The whole "complex" lies in a flat plane, and this is the same flat plane that the galaxies of the Local Super-cluster lie in. As yet, both observers and theorists are sceptical about Tully's claims. The Pisces-Cetus complex may be no more than a random pattern of galaxies and clusters in space—after all, the clusters have to be *somewhere*. The occurrence of superclusters and voids is, however, now beyond question. They represent the largest structures definitely identified in the Universe, geography on the grand scale. But on still larger scales—the largest observable scales—things do smooth out. Counts of the numbers of very high redshift galaxies seen in different parts of the sky show that the Universe is uniform on the grandest scale of all. Theorists seeking to explain how galaxies come to be here at all not only have to explain how individual galaxies form in those gravitational potholes, but why galaxies congregate in sheets and filaments in this way, with dark voids in between. One of the key puzzles is whether the dark matter in the Universe is distributed in the same way as the superclusters, so that the voids really are empty, or whether galaxies containing bright stars form only in special places, so that the voids may contain plenty of dark matter even if they contain few bright galaxies. The weight of evidence today is that bright galaxies are *not* good "tracers" of the way mass is distributed across the Universe, and that the Universe is a much smoother place than the pattern of galaxies on the sky would suggest.

A Background of Smoothness

Radio telescopes operating in the microwave band, sensitive to electromagnetic radiation with wavelengths of a few centimetres or less, detect a faint hiss of radio noise coming from all directions in space. This is the famous cosmic background radiation; even intergalactic space is not completely cold, but is filled with the dilute remnant, or "echo," of the hot radiation from the early phases of cosmic expansion. For the first few minutes, the temperature of this radiation exceeded a billion Kelvin (10^9 K)—hot enough for rapid nuclear reactions to occur. Such conditions can be described by equations tried and tested in the present-day Universe— they are used to calculate the workings of stars and of nuclear bombs. For the first microsecond, temperatures and energies were so high that we have less confidence in our understanding of the applicable physics. Throughout its first 10,000 years, the Universe must have been an opaque fireball. Then, the conditions everywhere resembled those in the *centre* of the Sun today. The radiation that radio astronomers detect as the cosmic background has propagated freely from the slightly later epoch when the whole Universe had cooled roughly to the temperature of the surface of the Sun today, a few thousand degrees Celsius. Until that time, the entire expanding Universe was still so hot that any electron that attached itself to a positively charged nucleus, such as a proton, would promptly have been knocked off again.

But at a critical moment in the evolution of the Universe, about half a million years after the moment of creation (or, looking back from our present perspective, at a redshift of about 1,000), the Universe cooled to the point where atoms could form and stay formed. From that moment on, essentially all of the electrically charged particles in the Universe, the electrons and protons, were locked up in stable, electrically neutral atoms. Because electromagnetic radiation cannot interact directly with neutral particles, but only with particles that carry electrical charge, from that moment on matter and radiation went their separate ways, and had very little more to do with each other. They "decoupled," and ever since that radiation has been cooling as the expansion of the Universe has redshifted its waves. Now, it is at a temperature of –270 degrees C, or 3 K—but it still pervades all of space; it fills the Universe and has nowhere else to go.

The exact temperature of the cosmic background radiation (actually a fraction less than 3 K) tells cosmologists about conditions in the early Universe and bears out the validity of calculations of the Big Bang based on general relativity. From our present point of view, however, what matters is not the precise temperature of this radiation but the fact that this temperature is exactly the same in *all* directions. Because the radiation has not interacted with matter since the time corresponding to $z = 1,000$, this tells us that the Universe was *extremely* smooth and uniform half a million years after the Big Bang. At that time, just before electrons and protons fused into atoms, they were still strongly coupled to the radiation that filled the Universe. So these measurements of the background radiation also tell us that the distribution of baryons at the time of decoupling was extremely smooth and uniform.

The most accurate measurements of the uniformity of the microwave background involve comparing the temperature of the radiation coming from different parts of the sky to an accuracy better than a thousandth of a degree. They show that patches of the sky that are each a few minutes of arc across (the angular size, viewed from Earth, of a protocluster of galaxies at redshift $z = 1,000$) all have the same temperature to within 1 part in 20,000. When the Universe was compressed by a factor 1,000 in linear dimensions, its average density was a billion times greater than it is now—far higher than the present-day average density even within a galaxy. Galaxies could not, therefore, have then existed as separate entities and it should not surprise us that the Universe was less "structured" at early times. But it is still surprising that we detect no hints of any irregularities over the sky due to "embryo" galaxies and clusters, which must already have been present at that stage. The question then arises of how quickly the structure can emerge from amorphous beginnings—are the conspicuous galaxies and clusters we now see around us compatible with such a smooth fireball?

Galaxies and clusters could condense from much less extreme initial fluctuations. A region that started off *slightly* denser than average, or expanding slightly slower than average, would lag further and further behind the rest of the Universe (so the contrast between the density inside the region and the density outside would steadily increase) until its expansion eventually halted and it formed a system held together by its own gravity. The density contrast grows in proportion to the scale of the Universe, and no faster. So any object, such as a cluster of galaxies, that has now stopped expanding must have been "overdense" by at least 1 part in 1,000 at $z = 1,000$. Since individual galaxies probably formed at a redshift of 5, the overdensity of an embryonic galaxy at $z = 1,000$ must already have been 1 part in 200. If the Universe was as uniform as the isotropy of the background radiation implies at a redshift of 1,000, but contained only the matter we can see today in the form of bright galaxies, irregularities as large as the superclusters we see today could never have grown as big as they are now.

This dilemma can be eased if there is dark matter in the Universe, in either of two ways. First, there might be a lot more matter filling in the gaps between the visible clusters and superclusters, distributed more smoothly than the distribution of galaxies, so that the density contrast today between a cluster of galaxies and a dark void is smaller than it seems. Some of this dark matter could be ordinary atoms—baryonic matter— like the stuff stars are made of. The second alternative is that the dark matter could have been distributed *less* smoothly than the baryons at $z = 1,000$. That is possible only if it is *not* baryonic matter and carries no electrical charge. Before that time, the Universe was dominated by radiation, and this sea of radiant energy would have prevented baryons from clumping together. Electrically neutral dark matter that did not interact

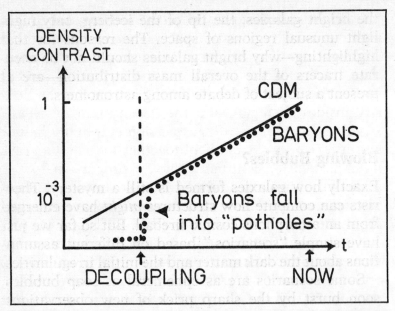

Figure 2.9 This graph shows how the density contrast of perturbations grows during the expansion of a "flat" Universe. If pressure forces can be neglected, the growth factor is proportional to the expansion and amounts to 1,000 since the epoch when baryons decoupled from the radiation. Radiation pressure prevents *baryonic* fluctuations from growing before that time; fluctuations in "cold dark matter" (CDM), however, are not inhibited in this way. The CDM therefore has a head start, and can create potential wells into which the baryons condense after decoupling. If the Universe is dominated by CDM, we can more readily reconcile the present "lumpiness" of the Universe with the apparently smooth baryon distribution at decoupling indicated by the microwave background isotropy.

with radiation, however, could have begun the segregation process and already produced regions of overdensity by the time of decoupling at $z = 1,000$. Such regions would not show up in the background radiation but would give the baryons a head start on the road to galaxy formation once they had formed neutral atoms and were no longer influenced by the radiation.

Either way, the background radiation is telling us that galaxies, and ourselves, could not have evolved without the help of dark matter. The dark matter would pervade the voids between the superclusters, even though the bright galaxies, the tip of the iceberg, only highlight unusual regions of space. The reasons for that highlighting—why bright galaxies should not be accurate tracers of the overall mass distribution—are at present a subject of debate among astronomers.

Blowing Bubbles?

Exactly how galaxies formed is still a mystery. Theorists can compute how structures *might* have emerged from an amorphous cosmic fireball. But so far we just have simple "scenarios," based on different assumptions about the dark matter and the initial irregularities.

Some scenarios are as ephemeral as soap bubbles, soon burst by the sharp prick of new observations. Sometimes, progress is made by combining the best features of several scenarios into a new model. But the most powerful insights come when different scenarios based on different assumptions, all point to the same requirement. One key feature of galaxy-formation scenarios based on up-to-date observations of the Universe is that they all use the simple laws of physics that have been developed from experiment and observation here on Earth. There is no evidence that any "new" laws are needed in order to explain how things came to be as they are. But it does seem clear that there is more to the Universe than the bright stars and galaxies.

Copernicus dethroned the Earth from a central position in the Universe; the cosmologists of the 1920s and 1930s demoted us from *any* privileged location in space, and taught us that the Milky Way Galaxy is simply an ordinary collection of stars located in an ordinary, typical region of the Universe. But now even "particle chauvinism" may have to be abandoned. The protons, neutrons, and electrons of which we and the entire astronomical world are made could be a kind of afterthought in a universe where totally different kinds of particles control the overall dynamics. Any hopes of quick progress towards understanding the stuff of the Universe may be quenched when we note that according to different scenarios (discussed later) 90 percent of the Universe is in the form of entities whose individual masses may range from 10^{-32} grams to 10^{39} grams—an "uncertainty" of more than 70 powers of 10. (Astrophysics may not always be an exact science, but seldom is the uncertainty as gross as this!) The uncertainty, however, lies in the choice between different scenarios, not within the scenarios themselves. By narrowing down our choice of options, we can still paint our pictures of how the Universe *might* have got to be the way it is, and search for the underlying deep truths and cosmic coincidences that link so many of the scenarios together.

CHAPTER THREE

★

Two Kinds of
Dark Matter

THE SIZES OF STARS, planets, and people are, as we saw in chapter 1, inevitable consequences of the relative strengths of the basic forces and constants of nature. Can galaxies and clusters of galaxies, the tracers by which we study the geography of the Universe, fit into the same basic scheme? One piece of evidence that suggests that some kind of cosmic rule must govern the size of galaxies, as well as the size of ourselves, is provided when we plot a graph showing how the mass of each kind of known object is related to its size. Figure 3.1 makes the point.

If there were no simple rule governing the sizes of things, then the points on this graph would be scattered all over the place. The fact that they are not reveals order in the Universe, and the nature of the regular distribution of the points, more or less along a straight line, gives a clue to what that order is.

First, there are two "forbidden zones" on the left of the diagram. To the upper left, there is a region corresponding to black holes, objects that have either a very large mass or a very large density, or both. The gravitational field of a black hole is so strong that nothing, not

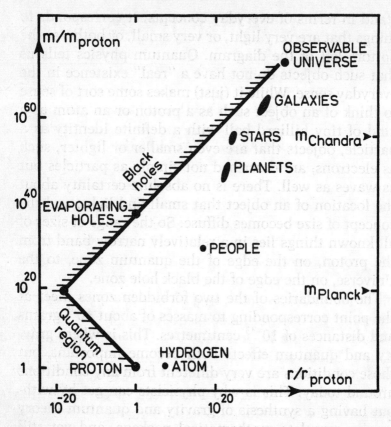

Figure 3.1 The characteristic masses and radii of various objects—from atoms to the entire Universe—plotted on a single diagram (on a logarithmic scale).

even light, can escape from it—so it is no surprise that we cannot see any objects in the Universe with sizes and masses corresponding to this part of the diagram. What is very interesting, and surely significant, is that the Universe itself sits just on the dividing line between this forbidden zone and the region occupied by stars, planets, galaxies, and ourselves (this is another manifestation of the flatness of the Universe). The whole Universe could, indeed, resemble a gigantic black hole, with spacetime bent around on itself.

The other forbidden zone is a little harder to understand in terms of everyday concepts. It corresponds to things that are very light, or very small, or both—at the bottom left of the diagram. Quantum physics tells us that such objects do not have a "real" existence in the everyday sense. While it (just) makes some sort of sense to think of an object such as a proton or an atom as a kind of tiny billiard ball with a definite identity as a particle, objects that are even smaller or lighter, such as electrons, are described not simply as particles but as waves as well. There is no absolute certainty about the location of an object that small and light, and the concept of size becomes diffuse. So the range of sizes of all known things lies in a relatively narrow band from the proton, on the edge of the quantum zone, to the Universe, on the edge of the black hole zone.

The boundaries of the two forbidden zones meet at the point corresponding to masses of about 10^{-5} grams and distances of 10^{-33} centimetres. This is where gravity and quantum effects both become important, but those conditions are very different from any conditions around today. This is why physicists can get by without having a synthesis of gravity and quantum theory in one complete mathematical package, and yet still come up with a good description of the way things behave in the Universe. *Either* gravity *or* quantum effects may be important for different objects in the Universe today, but never the two together. There has been no overlap between the gravity and quantum zones, *except* in the Big Bang.

Sizing Up Galaxies

We have already looked at the way a balance between electrical and gravitational forces keeps planets, stars, and ourselves on this line. But what are the balances— the cosmic coincidences—that restrict galaxies and clusters to well-defined parts of the diagram? The restrictions are not so clear-cut as they are for stars or ourselves, but there is a clear hint that even on the greatest scale the sizes of things are determined by the same basic laws of physics.

For the moment, we will ignore the dark matter that makes up 90 percent of the mass of the Universe, and puzzle only over the bright galaxies. Galaxies formed from huge clouds of gas, held together by gravity, which fragmented into stars. The type of galaxy that emerged (disc or elliptical) depends on details of the process— and, in particular, on how rapidly and efficiently stars formed while the protogalaxy was contracting. Even though there are many varieties of galaxy, there is no problem in identifying rough typical dimensions: around 10^{11} (a hundred billion) stars in a radius of 10 kiloparsecs (30,000 light-years). Is there a straightforward reason why the large-scale cosmic scene should be dominated by entities with this characteristic size and mass, just as there are physical reasons for the natural scale of stars? One suggestive physical argument has been taken seriously in the past few years and could offer a partial answer.

Imagine a collection of gas spheres, each held together by its own gravity. Such a sphere could be in equilibrium if it were hot enough for pressure to balance self-gravitation. The required temperature depends on just the mass and radius of each sphere (in fact, on mass divided by radius; this is a version of the two-thirds–power law mentioned in chapter 1) and is easy to work out. Any hot gas loses energy by radiating it away into its surroundings; the radiation rate, which depends on how hot and dense the gas is, is also easy to calculate. If a cloud radiates *slowly*, it will gradually deflate but stay in equilibrium as a single homogenous mass. On the other hand, if the cooling proceeds too fast, the cloud cannot retain its pressure support. It would then collapse in a free fall and fragment into smaller pieces. Any protogalaxy in which stars can form must be in this second, "fast cooling," regime.

It isn't difficult to calculate the demarcation between the two regimes and discover which masses and sizes of clouds could fragment. This depends on basic physical constants—the strength of gravity and the atomic constants that control how much radiation a hot gas emits. It turns out that fragmentation occurs for any clouds less massive than 10^{12} times the mass of our Sun and with radii below 75 kiloparsecs—but not for heavier and larger clouds. These critical dimensions (which few physicists would have guessed beforehand, even to within a factor of a million) are similar to those of the biggest galaxies. The "cooling versus collapse" argument plays a role in most detailed cosmogenic schemes; it offers a convincing explanation of why galaxies cannot be even bigger than they are.

One amusing result of this—not really a coincidence, since it depends on simple rules of physics—relates to the geometric means of the sizes of important features of our Universe. The geometric mean is obtained by multiplying the lengths of two things (or their diameters) together and taking the square root of the product; it is a more useful way of averaging things with very different sizes than the arithmetic mean we use in everyday life. The size of a human being is the geometric mean of the size of a planet and the size of an atom; the size of a planet is the geometric mean of the size of an atom and the size of the Universe. Both, like the sizes of galaxies, are a result of the balance between gravitational and electrical forces.

It is, however, unreasonable to expect that galaxies can be explained as straightforwardly as stars. After all, we can see stars forming now, close at hand in the Milky Way. Galaxies formed during a remote cosmic epoch. Any understanding of galaxies must entail placing them in a cosmological context; physical processes can perhaps select the appropriate range of masses and radii, but only if the cosmologists can give us a variety of protogalactic clouds in the first place.

The Universal "thermal soup" of the Big Bang started off almost featureless, but not quite. There must have been (we don't know why) small initial fluctuations from place to place in the expansion rate or the density. Structures would then have emerged, as overdense regions lagged more and more behind the Universal expansion, and eventually halted and formed systems bound together by gravity.

Theorists trace the history of the Big Bang right back to the so-called Planck time (10^{-43} seconds), smaller by about 60 powers of 10 (60 "decades") than the Universe's present age. Many key features of the Universe, including the initial fluctuations, were imprinted during the very early stages. Because the physics of the ultracompressed, high-density stages is speculative, we have no firm understanding of exactly where the fluctuations came from. This is related to the problem of why, on the largest scales, the Universe is so uniform and flat. We shall return to this issue later, but for the moment it must be regarded as a coincidence that the initial fluctuations were adequate to initiate galaxy formation, without being so large that the Universe ended up in a chaotic mess.

Physicists have identified various phases in the expanding Universe where crucial processes, or transitions, would have occurred. Without engaging in technicalities, we can divide cosmic history into three parts. For the first 10^{-4} seconds (the first 40 "decades" of logarithmic time) everything was squeezed to supernuclear densities, and particles had such high energies that the relevant physics is not merely unfamiliar but in some regards completely unknown (or, at best, controversial). After this time, the microphysics becomes less

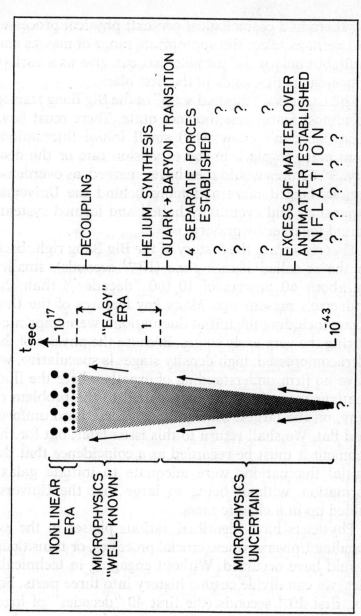

Figure 3.2 Key events in the history of a "hot Big Bang" universe.

exotic—less remote from conditions created in laboratories on Earth. During this era after 10^{-4} seconds, nuclear reactions would have occurred. The rates at which these reactions took place, and the abundances of the elements produced in those reactions, depend on how densely packed the baryons were when the temperature was 10^9–10^{10} K. Since we know the present temperature of the Universe (2.7 K), we can relate the element abundances to the present-day density of baryons in the Universe. Confidence in the Big Bang theory was boosted by the gratifying agreement between the observed abundances of helium and deuterium in stars and these theoretical predictions. Indeed, cosmologists use these calculations to *infer* the density of baryons from the measured element abundances; the evidence about the first few minutes of the Universe, provided by these relic chemical elements, is no less firm than some of the inferences geologists and paleontologists make about the early history of Earth.

The second, and "theoretically easy," era of cosmic history continues until the overdense regions, which have become even more distinct from their surroundings because gravity gives them an above-average deceleration, start to condense out and collapse. After this stage, discrete objects would exist, and astronomers could in principle observe them. But theorists are then faced with a new set of difficulties. The essential physics is just Newtonian gravity and gas dynamics; but the complications are those of "nonlinearity," and the phenomena become hard to understand for the same reasons as, for example, weather prediction is difficult—each little bit of the system obeys simple physical laws, but vast numbers of little bits interact with one another in complex ways.

A key cosmogenic question is: How much of galaxy formation can, even in principle, be explained by processes occurring at the relatively recent epochs accessi-

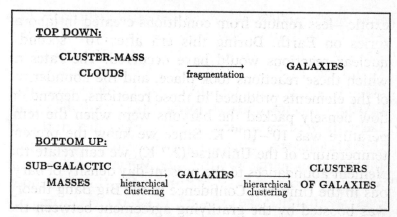

Figure 3.3 The contrast between a "bottom up" cosmogenic scheme, where small systems condense first and then cluster hierarchically, and a "top down" scheme, where the first systems to condense have the mass of an entire cluster of galaxies and subsequently fragment. A Universe dominated by cold dark matter (CDM) exemplifies the first option. If, however, hot dark matter (e.g., fast-moving neutrinos) were dynamically dominant, then fluctuations on all scales smaller than a cluster of galaxies would be homogenised, and galaxies would form after clusters.

ble to observation, and how much has to be attributed to features imprinted at earlier times? There is no obvious feature of the early Universe that would "know" the scale of galaxies in advance, so we would expect the initial fluctuations to have a smooth "spectrum," with all kinds of variations on many different scales. If all the fluctuations, on all scales, grew by gravitational amplification as the Universe expanded, the relative amplitudes of fluctuations on different scales would stay the same, just as set up by the initial conditions. But there are other processes at work, connected with pressure forces and the random diffusion of particles, which can selectively damp down some fluctuations and amplify others; these are also fairly straightforward to analyse during the "theoretically easy" era, when the amplitudes are still small. Although the Universe as a whole expands, overdense irregularities expand more slowly, and eventually reach a maximum size before beginning to fall back upon themselves as their own gravity overcomes the expansion of the Universe. In a so-called bottom-up scenario, objects like dwarf galaxies and globular clusters formed first and the pieces were then grouped together by gravity to make galaxies and clusters. In the alternative top-down scenario, clusters or superclusters formed first and then fragmented. Neither picture gives a perfect description of the real Universe, but each provides insights into how things got to be the way they are. When we try to paint a slightly more realistic picture of the Universe, however, we find that the existence of galaxies is intimately connected with the presence of dark matter.

A Biased View

The distribution of galaxies across the sky is just about the only piece of evidence we have about the geography of the Universe at large. The total amount of mass associated with galaxies can best be estimated by studying the distribution and motion of bright galaxies in clusters. Studies of the spectra of these galaxies show, by the Doppler effect, how they are moving. Knowing how fast the galaxies in a cluster are moving relative to one another, we can estimate the total mass needed to stop the cluster from flying apart and make it gravitationally bound. Such studies corroborate the evidence from studies of individual galaxies: there is about ten times as much matter in some dark form as in the stars and gas we see. But the average smoothed-out density of all this dynamically inferred matter is still no more than 10 to 20 percent of the "critical" density needed to make the Universe flat. Curiously, detailed calculations of element production during the Big Bang also say that the density of baryons (protons plus neutrons) in the fireball is no more than 20 percent of what would be required for the Universe to be flat. This may

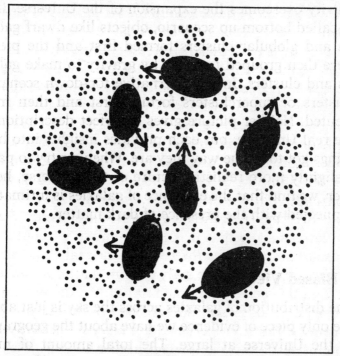

Figure 3.4 The random motions of the galaxies in a cluster can be inferred from the Doppler effect. To stop the cluster from flying apart, it must contain ten times as much dark matter as the aggregate mass observed in all its constituent galaxies.

be a genuine coincidence, not one of the cosmic coincidences we are interested in, and one that misled cosmologists for twenty years into accepting that baryons (some in bright galaxies but 90 percent in some dark form) constitute all the matter there is in the Universe. Most cosmologists now rate the theoretical case for a flat Universe as quite compelling, however, and the only other piece of evidence we have, the uniformity of the background radiation, supports that view. Motivated as much by theoretical prejudice as by observational evidence, they have become sceptical as to whether bright galaxies really do tell the whole story about the way matter is distributed in the Universe.

Big "voids" in the distribution of bright galaxies are

now known to be quite common. Within voids, bright galaxies are at least ten times more sparsely distributed than across the Universe at large. One way of trying to understand how such features can emerge in the real Universe is with the aid of computer simulations. Starting out with a distribution of galaxies, long ago, that matches the requirements of the measurements of the background radiation, the computer models are allowed to "evolve" with the points representing galaxies clumping together under the influence of their mutual gravitational attraction, while the clumps themselves drift apart as the model universe expands. Such simulations show that it is impossible to evacuate the voids as completely as the voids in the real Universe. If the Universe contains exactly enough matter to make it flat, and this is all distributed initially in the same way as bright galaxies are, no void can have less than 25 percent of the average density today. If the Universe contains less mass, it is even harder to evacuate the voids. Mass must "hide" in the voids to escape detection, while bright galaxies provide a biased view of the Universe.

There is no rule of nature that says all of the dark matter must be the same stuff, and it would be surprising if it were. But the simplest way to begin to understand what the dark matter might be is to sketch out scenarios that are each based on the assumption that only one kind dominates. The key point is that the dark matter that dominates the dynamics of a flat universe *cannot* be the kind of matter (baryonic matter) found in stars, planets, and ourselves, since the maximum amount that could have emerged from the Big Bang, if our ideas on element production are correct, falls short of what is needed by at least a factor of 5. If there is indeed only one important kind of dark matter, the flatness of the Universe tells us that it is less clumped than galaxies, that voids are not as empty as they look.

There are two principal ways in which this could have come about.

First, the baryons themselves could be segregated from the dominant dark matter, all the baryons being in the bright galaxies that form the sheets and filaments we can see, while most of the dark matter is in the voids. Alternatively, the baryons and the dark matter may be similarly distributed in the Universe at large so that the two forms of matter intermingle even in the voids. We would then see bright galaxies distributed in bubbles and filaments around those voids because some special conditions are needed to trigger the creation of clusters of bright galaxies, and those special conditions simply do not occur everywhere. Baryons may be present in the voids, but they have failed to light up and display their presence to us.

One specific possibility, the front-runner among such theories today, is that bright galaxies formed *only* in regions of space where the density of the matter in the Universe was *exceptionally* high. You can think of this in terms of waves on the ocean, or the high peaks of a mountain range. Only the highest wave crests, or tallest mountain peaks, represent galaxies; the troughs of the waves, and the valleys of the mountain range, represent the dark voids, which still contain an enormous amount of matter (water or rock, depending on which image you prefer). If the density of the Universe varies from place to place in a random manner, with minor fluctuations occurring within larger fluctuations (like waves on top of a long ocean swell), this would provide a very natural way to produce clusters and superclusters of galaxies, with voids in between. In a region where a large fluctuation (a swell) has increased the density already, places where there is a small extra fluctuation (a wave) that increases density a little more will form galaxies. In a region where there is a correspondingly large reduction in average density (the trough

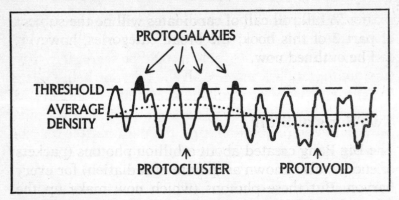

Figure 3.5 If galaxies form only from exceptionally high peaks in the initial density distribution, they will be strongly concentrated in the crests rather than the troughs of long-wave perturbations. They will therefore form more readily in incipient clusters and be deficient in incipient voids. The galaxies will consequently be more "clumpy" than the overall distribution of mass.

of the swell), small-scale fluctuations (waves) that increase density a little can never become dense enough to form galaxies.

Furthermore, once bright galaxies do form in a developing cluster or supercluster, their very presence may trigger the formation of more galaxies in the high-density region in which they are embedded. If galaxy formation spread like an epidemic, galaxies would naturally end up in clusters, and large regions (the voids) might escape "infection" completely.

Rather than this "biasing" being some kind of desperate last resort by theorists trying to reconcile the geography of the Universe indicated by studies of bright galaxies with the requirement that the Universe is flat, it would be astonishing if no such effects were at work; indeed, several different biasing processes might each have played a role when the Universe was young, producing a cumulative effect. The notion that galaxies trace mass is an unjustified assumption, and finding the actual bias mechanism is a problem that is intimately related to finding the actual nature of the dark

matter. A full roll call of candidates will be the subject
of part 2 of this book; the broad categories, however,
can be outlined now.

Two Sorts of Stuff

The Big Bang created about a billion photons (packets
of energy, also known as quanta of radiation) for every
baryon. But these photons (which now make up the
microwave background) have zero *rest* mass, and the
mass-equivalent of the energy each one carries (calculated from Einstein's $E = mc^2$) is so small that, in spite
of their overwhelming numbers, they only contribute
one ten-thousandth of the actual cosmological density.
It is a firm prediction of the Big Bang theory that there
should also be a "neutrino background"; about as many
neutrinos are produced in the Big Bang as there are
photons. Neutrinos must therefore, like photons, outnumber baryons by a factor of about 10^9; so their overall gravitational influence on the dynamics of the
Universe will be important even if their individual
masses are tiny compared with those of the more familiar
elementary particles such as electrons and protons. The
electron, which is about the lightest thing that has any
direct influence on our daily lives, has a mass of about
500,000 electron Volts, or eV. This is roughly 10^{-30}
kilograms, but it isn't really helpful to try to relate eVs
to everyday units; nobody can get a real "feel" for the
mass of an electron. What matters is that if neutrinos
provide enough dark stuff to flatten the Universe then
each would weigh a few tens of electron Volts—less·
than one ten-thousandth of the mass of an electron.

Because they are very light, such particles are born,
in the fireball of the Big Bang or in nuclear reactions
inside stars today, travelling at very nearly the speed of
light. It is very difficult for particles travelling so rap-

idly to be bound together into clumps by gravitational bonds, and in the early stages of the expanding Universe they would have streamed in all directions very smoothly and uniformly, homogenising their distribution throughout space; their presence would also have tended to smooth out all small-scale irregularities in the distribution of baryonic matter, just as the tide rushing in over a beach smooths out the footprints left in the sand, even though each grain of sand is much more massive than a molecule of water. As the Universe expanded and cooled, neutrinos would have been spread more thinly, and they would have slowed down from their initial superhigh speeds. Eventually, they would be moving slowly enough to allow irregularities to begin to grow by gravitational attraction, and at that point the first structures would form. But those structures would not be on the scale of globular clusters, galaxies, or even clusters of galaxies, because the neutrinos would have been homogenised on these relatively small scales. The *first* structures to appear in a neutrino-dominated universe are on the scale of superclusters, shaped like huge sheets and filaments, and wrapped around enormous voids in which no gravitational condensations occur.

For a while, this made the neutrino-dominated scenario attractive to cosmologists. But it soon ran into serious difficulties. In such a top-down scenario, superclusters break up into clusters, which break up into galaxies, which break up into stars. All this takes time, and computer simulations show that superclusters would not form before $z = 3$. Galaxies (which, in this scheme, cannot form until later) would then emerge only very recently, at a redshift much less than 3. This is hard to reconcile with the discovery of quasars at redshifts greater than 4, since quasars are thought to be the active cores of young galaxies. There are other problems. Why, if galaxies form so late, do the oldest stars in our Galaxy, notably the stars of the globular clusters, seem to be nearly as old as the Universe itself? And where do structures as small as globular clusters and dwarf galaxies come from? Neutrinos with the right mass range cannot form gravitational condensations that small *at all*, if our understanding of the Big Bang and the laws of physics is correct.

Most of these difficulties would be resolved if the dominant dark matter consisted of particles that were "cold," in the sense that they had low random speeds and therefore did not disperse and homogenise on galactic scales, as would neutrinos (which are, by contrast, described as "hot" dark matter). The distinction is like the difference between molecules of liquid water, which are cold and do not move very fast, and molecules of water vapour, which move faster because they are hotter. But the analogy is not exact, because in cosmology "hot" particles are not simply "cold" particles that carry more energy, they are a different family of particles altogether. That is why, ten years ago, the proposal that the Universe might be dominated gravitationally by cold, dark matter was a rather daring proposal. The cold matter could *not* simply be neutrinos that had run out of steam; it had to be something new entirely. Cosmologists can define the properties dark stuff "ought" to have in order to explain observed features of the Universe, but no particles with those properties were known at the time. Indeed, none are *known* to exist even today. But at around the same time, and throughout the 1980s, particle physicists studying interactions at high energies in accelerators here on Earth (such as the machines at CERN, the European Centre for Nuclear Research, in Switzerland, and Fermilab in the United States) were proposing the existence of "new" particles to plug gaps in their theories. Several of the particles required by those theories have exactly the right kind of properties to make them suitable candidates for cold dark matter particles in a flat universe. The disadvantage is that their existence is not proven; the encouraging sign is that the same kinds of particles are required to explain observations at opposite ends of the spectrum of science, in the Universe at large and on scales smaller than an atom.

This is another remarkable coincidence, though of a slightly different kind from the ones we have been discussing so far. It is worth emphasising the point. Particle theorists, trying to develop a complete description of the world of the very small, are forced to postulate *exactly* the kind of particles that cosmologists, contemplating the world of the very large, need to explain the structure of the Universe.

So what is cold dark matter (CDM) and what happens in a universe that contains enough of it to be flat? The important thing about CDM is that the particles move slowly, much slower than the speed of light. Slow motions would automatically be expected if the individual particles were heavy, compared with, say, an electron. Some candidates are particles with several times the mass of a proton, which is itself almost 1 billion electron Volts (1 GeV), 1,840 times the mass of an electron. (One CDM candidate, however, has a very light mass, like the neutrino, but is born with low velocity. This is the axion, described in chapter 4.) The low velocity of all CDM particles means that they can be bound together by gravity more easily than, for instance, "hot" neutrinos. Where the dark matter forms a clump, baryons have to follow, tugged inward by the gravity of the dark stuff, like water flooding into a pothole in the road.

Irregularities in the density of CDM can begin to grow much sooner after the Big Bang. We have a bottom-up scenario, in which there is no problem explaining how features as small as dwarf galaxies could form, and in which galaxies group together to form clusters, which group into superclusters and so on. On the scale of galaxies, CDM can account very nicely for details of the structure and shape of a galaxy like our own, and the way in which it rotates. On the scale of superclusters, provided the right amount of biasing is invoked (which has to be done in any scenario, not just with CDM), numerical simulations show that the galaxies do indeed group into filaments and sheets surrounding dark voids.

A universe dominated by hot neutrinos is predicted to have a rather simple structure, like the cells of a honeycomb (though not so regular), in which bright galaxies form only in well-defined sheets and not at all in the voids. The CDM universe is more messy and complicated, with a richer structure that perhaps looks more like the real Universe. Sheets and filaments do form, but they intertwine in a complicated way; and although bright galaxies form preferentially in the filaments, as a result of biasing (by whatever means), there is every reason to expect to find in the voids fainter galaxies, dwarf galaxies, and even huge clouds of gas that failed to complete their collapse and breakup into galaxies and stars. Probably the single key observational discovery (apart from finding CDM particles in the laboratory) that would tilt the balance conclusively in favour of the CDM scenario would be the discovery of faint galaxies in the voids; if they exist, this will be possible when NASA's Hubble Space Telescope eventually is put into orbit.

We should not pretend that CDM scenarios have now swept the board. Some theorists, for example, still claim that neutrinos with a very small mass could provide the dominant form of matter that determines the way galaxies form. The more successful neutrino scenarios, however, now involve masses that are rather too small for neutrinos to provide all the dark matter needed to flatten the Universe, and these low masses are also

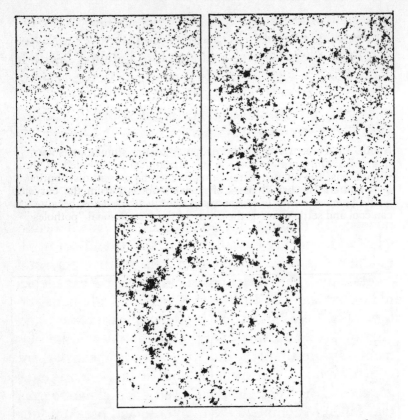

Figure 3.6 This diagram (along with the following two) shows the results of computer simulations of gravitational clustering in the expanding Universe. The contents of a typical cubical volume are computed. The three panels show three epochs (time increasing from left to right), the panels having been rescaled to allow for the overall cosmic expansion. For initial conditions similar to those for cold dark matter, the mass-scale of clustering grows in a hierarchical fashion. (Simulations courtesy of M. Davies, G. Efstathiou, C. Frenk, and S. White.)

consistent with limits set by the latest experiments and observations. Of course, it may be that there is more than one kind of dark stuff in the Universe, and that their combination of masses and densities conspires to make the Universe flat. But a universe with about ten times as much cold dark matter as baryonic matter has properties that are remarkably similar to those of the observed Universe. It seems to be the best simple model

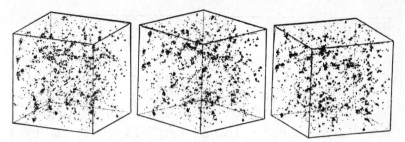

Figure 3.7 Three views of the endpoint of the simulation shown in figure 3.6. Only gravitation (no gas dynamics) is included in these simulations. The picture therefore represents the present distribution of *dark* matter. The bright parts of galaxies (formed from the baryons) would be more centrally concentrated than the dark halos, because gas can cool and settle towards the centre of the gravitational "potholes."

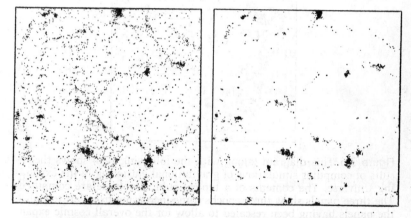

Figure 3.8 This shows the endpoint of a simulation for *hot* dark matter. Note the evident filamentary structure. The left-hand panel shows the overall mass distribution; on the right, the expected distribution of bright galaxies is depicted.

available, and well worth closer scrutiny and testing before we introduce any complications of this kind, only when (and if) they prove essential. We need more information on how galaxies are distributed, how they are moving, and a clearer theoretical picture of how the luminous parts of galaxies form within the dark haloes.

Before the Galaxies

How large were the first pregalactic objects? And when did they form? When the Universe was about a million years old, and cooled below a temperature of a few thousand degrees, the fireball radiation redshifted from the visible band to the infrared. The Universe then entered, quite literally, a "dark age," which continued until the first gravitationally bound objects condensed and "lit up." In a neutrino-dominated universe, the dark age would persist for a billion years, until huge clouds the size of superclusters collapsed and fragmented into galaxies. In a CDM-dominated universe, on the other hand, smaller-scale fluctuations would have formed bound systems *before* galaxies formed. The gravitational potholes due to bound clumps of CDM heavier than a million solar masses would be deep enough to overcome pressure forces in the primordial gas; in this kind of bottom-up cosmogony, the first stars would form from clouds with a million times the mass of our Sun.*

This is a very important possibility, because these objects, small compared with galaxies, could change the environment of the expanding universe before galaxies form. They could even be the "seeds" from which galaxies grow. We do not understand star formation well enough to be able to decide whether a gravitationally bound primordial cloud containing a million solar masses of material will form a single supermassive star or fragment into a cluster of more ordinary stars. Supermassive "stars" run through their life cycles very

---

*These objects should not be confused with globular clusters, which have similar masses but must have formed later. Interestingly, however, recent observations have shown that the globular clusters in all galaxies are very similar, even though the galaxies in which they are found may be superficially very different from one another. Globular clusters form, it seems, as an inevitable by-product of the collapse of a cloud of baryons with galactic mass; in those circumstances, the mass scale of around a million Suns again arises naturally.

quickly and then explode, sending a blast wave through the surrounding gas, scattering elements built up by nuclear reactions, and, perhaps, leaving a massive black hole behind. It is easy to see how such objects could affect their surroundings; they could trigger enormous explosions, leading to biased galaxy formation. A massive black hole would be a gravitational pothole par excellence, sucking in matter to surround itself in a cloak of gas in which stars might form to make a galaxy. Explosions could generate galaxies if stars with a mass of "only" a hundred times the mass of the Sun existed by a redshift of 10, which seems well within the realm of possibility.

Could such pregalactic objects ever be detected by the traces they left on their surroundings? In principle, they could—and in practice, they may already have been found, by observations made on a rocket flight in 1987.

One effect of a burst of star formation when the Universe was very young would be, as we have mentioned, to lace the Universe with traces of elements built up by nuclear reactions inside those stars. Dust rich in silicon and carbon would have absorbed the light from these early stars. Like the surface of the Earth (or a car left parked in the Sun), which is warmed by the Sun and reradiates the energy it receives in the infrared part of the spectrum, this early cosmic dust would have got hot and reradiated its energy at wavelengths just shorter than the wavelength of the main peak of the energy of the background radiation itself. It would provide an extra component of background radiation, enough to show up even today, after the radiation has cooled all the way to 3 K.

Theorists who favour the early-star hypothesis were able to calculate in what waveband observers should look for the telltale signature of those stars, today. Frustratingly, the calculated position lies in a part of the microwave band, at submillimetre wavelengths, where

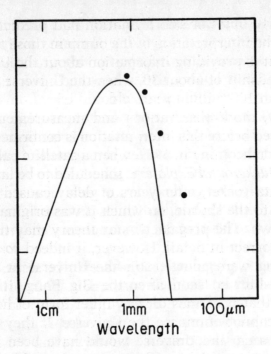

1cm   1mm   100µm

Wavelength

Figure 3.9 The spectrum of the background radiation at millimetre wavelengths: the dots show the intensity measurements obtained in a 1987 rocket flight by T. Matsumoto and his collaborators. The curve shows a black body at 2.74 degrees, the best fit to all other data at wavelengths longer than a millimetre. If this "submillimetre excess" is real, it suggests that there was a colossal energy input from Population III stars at redshifts exceeding 10, and that this starlight has been absorbed and re-emitted by dust.

observations from the ground are blinded by interference from radiation by molecules such as water in the Earth's atmosphere. Only a very sensitive instrument, flown in space, could make the necessary observations.

Just such an experiment was flown by a joint American-Japanese team in February 1987, in a rocket, launched from Japan, that nosed above the atmosphere for ten minutes. The instrument found exactly the kind of bump in the microwave background spectrum that the theory of an early burst of star formation had predicted. Assuming the interpretation of the bump in those terms is correct, it is providing information about the Universe from a redshift of about 30, when the Universe was no more than 100 million years old.

Clearly, more observations and measurements will be required before this interpretation is confirmed. They may be forthcoming in 1989, when a satellite called the *Cosmic Background Explorer* is scheduled to be launched on a Delta rocket, after years of delay caused by the accident to the shuttle, on which it was originally due to be flown. The pregalactic-star theory may turn out to be incorrect in detail. However, if indeed *something* interesting were going on in the Universe at such a high redshift, so soon after the Big Bang, it would indicate that neutrinos (or any other form of hot dark matter) cannot dominate the Universe. If they did, as we have seen, the Universe would have been far too smooth at these high redshifts to produce this kind of effect. Cold particles really do seem the best candidates for the bulk of the dark matter. But why should the balance between dark stuff and bright galaxies have fallen out the way it has?

Another Coincidence?

When we talk about the possibility that several different kinds of matter might be present in the Universe, including CDM (possibly in more than one variety), baryons, and perhaps even a little HDM, it might at first sight seem like a strange conspiracy that the masses of all these different things should add up to exactly the right amount to make the Universe flat. In fact, this is not a coincidence that should cause any sleepless nights. The flatness of the Universe would be imposed in the Big Bang itself, perhaps at the very moment of creation, guaranteeing the "right" amount of energy to be converted into mass, in line with Einstein's equation $E = mc^2$, to make the Universe flat. This is a requirement of, for example, the inflationary theories of the origin of the Universe (described in *In Search of the Big Bang*); from another perspective, it is an intriguing anthropic coincidence. But once that amount of energy was available to be turned into mass, it is no surprise to find that by adding up all the different masses it went into, we end up with the same amount of mass-energy we started with. If we had a gallon can full of water, and used it to fill a variety of different pots, pans, and bottles, it would hardly be an astonishing coincidence that the water in all those vessels put together was exactly enough to fill a gallon can to the brim.

But there is another way in which the distribution of the available matter looks peculiar. Why didn't all of the energy go into CDM particles, or all of it into baryons? Evidence that there is ten times more dark matter than baryons around in the Universe suggests a big difference between the two in everyday terms, but really the numbers are surprisingly close to each other. There could have been a million, or a billion, times more dark stuff than baryons, or the other way around. In either extreme case, it seems very unlikely that galaxies like the one in which we live could have formed.

We believe that the ratio of dark stuff to baryons, 10:1, may be a significant number in cosmology, and that if this number were very different we would not be here to puzzle over it. But, as yet, we can only regard it as an unexplained cosmic coincidence, since nobody has devised any convincing theory to explain why the primordial mass-energy should have been split in exactly this way. If it is ever understood, this coincidence will surely give a vital insight into the nature of physics at a fundamental level; meanwhile, we must be content with trying to understand, on the basis of the most promising current theories of particle physics, what the dark matter itself might be. But a little speculation on why this ratio should hold is surely in order.

The microwave background offers direct evidence that there was indeed an early fireball phase of the Universe, long before any galaxies existed. The cosmic helium abundance gives us some confidence that we can push our understanding of the evolution of the Universe back to a time when it was only one second old. But the question of why the Universe at $t = 1$ second was expanding in the "right" way and why it contained the particular mix of baryons, radiation, and dark matter that we infer still seems to demand an appeal to initial conditions. "Things are as they are because they were as they were." Our inferences come up against a barrier, just as did the ancient Indian cosmologists who envisaged the Earth supported by four elephants standing on a giant turtle, but did not know what held the turtle up.

The key features of the Universe were imprinted at times earlier than one second, and the earlier back we extrapolate, the less confidence we can have in the adequacy or applicability of known physics. Theorists differ in how far they are prepared to extrapolate back with a straight face. Some have higher credulity thresholds than others. But those whose habitat is the "gee whizz" fringe of particle physics are interested in the possibility that the early Universe may once have been at colossally high temperatures, because this provides the only testing ground for some of their theories. Particle accelerators on Earth are inadequate to provide an analog of these conditions.

The motivation of these theorists is linked with their search for a unified theory of all the forces of nature. The stumbling block is that the critical energy required to carry out tests of the predictions of their favoured theories is about 10^{15} giga electron volts (GeV). By comparison, the latest (successful) test of a theory of partial unification took place using particles accelerated to an energy of about 100 GeV at CERN. The next step requires energies 10 million million times higher still, a million million times greater than there is any hope of achieving in experiments on Earth. We have to extrapolate our inferences about the Big Bang all the way back to the first 10^{-35} seconds to find a time when particles were so energetic that they were colliding with one another at energies of around 10^{15} GeV. Perhaps the early Universe was the only accelerator where the requisite energy for unifying the forces could ever have been reached; however, this accelerator shut down 10 billions years ago, and we can learn nothing from its activities unless the era around 10^{-35} seconds left some fossil behind (just as the helium in the Universe is a fossil left over from the first few minutes). Physicists would seize enthusiastically at even the most trifling vestige surviving from that era. But it has actually left some very conspicuous traces indeed—it may be that all the atoms in the Universe are essentially a fossil from 10^{-35} seconds. Grand unified theories predict a slight excess of matter over antimatter emerging from this era, so that, when baryon-antibaryon pairs later annihilate into photons, one "surplus" baryon survives for every billion photons thereby created.

But what about the dark matter? The fireball must have created other species of particles, and their antiparticles, as well as baryons. All of these were in equilibrium in the early, dense phases, and present in roughly equal quantities. The ratio of baryonic to dark matter masses, 1:10, must have been decided at about the same time as the ratio of baryon to photon numbers, $1:10^9$; it could have happened in one of two ways. The neatest possibility would be if there were a similar favouritism for particles over antiparticles in the case of dark matter. Then, the required ratio would arise automatically if the mass of each dark matter particle was around ten times the mass of a proton—10 GeV instead of 1 GeV. There would be equal numbers of DM particles and baryons in the Universe today, with one set ten times heavier than the other.

If there were no such preference for matter over antimatter in the case of DM, you might expect that as the Universe cooled, all the DM particles would meet their DM antiparticle partners and annihilate, so that none would survive. However, it could be that the chance of such a particle and antiparticle interaction is small enough for some to survive into later eras. Then, as the Universe expanded, the DM particles (and antiparticles) would be spread ever more thinly across the Universe, with less and less chance of meeting each other and annihilating. The number of survivors could, in principle, be calculated using a complete theory, which would prescribe exactly what chance there was of a particle/antiparticle pair interacting under the conditions that prevailed in the fireball. Theorists are far from having such a complete description of events at these high energies, and at present it seems that if this is the way energy was shared out among baryons and dark matter, then it is a coincidence that there should be just ten times as much mass in dark matter as in baryons.

The first alternative seems much neater, to us, and we will be intrigued if any of the searches for dark matter particles now going on discover particles with masses around 10 GeV. But we won't be unduly surprised if this "prediction" fails. When dealing with phenomena such as ordinary stars, we feel fairly confident that we know the relevant physics. When conditions are more extreme, such as in the centres of galaxies, we are less confident, although it is astonishing how far we can go without running up against a contradiction.

Our confidence in forecasts based on inferences about the first 10^{-35} seconds is not high!

There is no agreed understanding of how the Universe combines the small-scale roughness needed to initiate galaxy formation with the overall uniformity that has allowed it to expand smoothly for more than 10 billion years. It seems that the early Universe was smooth only in the sense that the ocean is smooth, but we do not know what determined the small-scale roughness. There is no reason to expect that the "waves" would exist only on the scales of galaxies and clusters. It is more likely that they would spread over all scales.

The simplest and most natural assumption is that the Universe is equally rough on every scale. There is then a single pure number, the "roughness parameter," which characterises the fluctuations and must have been imprinted on the Universe at a very early stage. In the CDM model, a value for this number of around 10^{-5} accounts for both the properties of galactic haloes and the way galaxies are clustered. Moreover, irrespective of the detailed cosmogonic model, this basic number is still pinned down within a narrow range. A value much larger than 10^{-5} is ruled out by the uniformity of the microwave background temperature across the sky. If the number were much bigger, bound systems would also have condensed out early on—perhaps even during the fireball phase of the Universe when matter and radiation were still coupled together by electromagnetic interactions. That would have prevented fragmentation and ensured that the Universe became dominated by gigantic black holes, each one containing more mass than an entire cluster of galaxies. On the other hand, if the initial roughness were substantially less than 10^{-5}, the Universe would still today, after more than 10 billion years, be amorphously uniform, with no galaxies, no stars, and no life. This shows that the fact that the Universe is rough enough, but not too rough, for life to

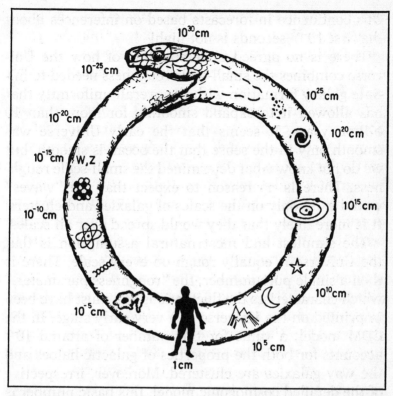

Figure 3.10 Most features of our everyday world are determined by the nature of atoms (10^{-8} cm); the properties of stars depend on the physics of atomic nuclei (10^{-13} cm). New ideas in particle physics suggest further linkages between micro and cosmic scales (i.e., between left and right in this picture): for instance, the dominant dark matter may be subnuclear particles surviving from the Big Bang. The ultimate unified theory—here symbolised "gastronomically"—may relate quantum gravitational effects (on scales of the Planck length, 10^{-33} cm) to the properties of the entire observable Universe (10^{28} cm).

have emerged depends on yet another cosmic coincidence, the value of the roughness parameter. The large-scale geography of the Universe is once again seen to be intimately connected with our own existence.

One theme that has emerged from recent research—and will be a recurrent theme throughout this book—is the interdependence of different phenomena, illustrated picturesquely in figure 3.10. The everyday world is determined by the physics of atomic nuclei, and the much larger structures (galaxies and clusters) may be gravitationally bound only because they are embedded in clouds of subatomic particles that are relics of the very early Universe.

In considering the earliest stages of the Big Bang, however, we confront conditions so extreme that we know for sure that we do *not* know enough physics. In particular, we lack a quantum theory of gravity. The two great foundations of twentieth-century physics are quantum theory and general relativity, but there is no overlap between the two today. Quantum effects are crucial on the submicroscopic level of elementary particles, but there gravity is too weak to play a part; gravitational effects dominate on the scale of planets, stars, galaxies, and the whole Universe, but on such a scale quantum effects, such as uncertainty, can be ignored. But when the Universe was squeezed to colossal densities and temperatures, gravity could be important on the scale of a single particle. This happened at 10^{-43} seconds, the Planck time. Before then, the effects of quantum gravity would be dominant, and since we have no quantum theory of gravity, even the boldest physicist can look no further back into the history of the Universe.

All our thoughts about the initial instants of cosmic history—grand unified theories and the rest—are still tentative. But these studies at least bring a new set of questions—the origin of matter, for instance—into the scope of serious discussion. The realisation that protons do not last forever, another inference from grand unification, suggests, moreover, that the Universe may possess no conserved quantities other than those, such as electrical charge, which average out precisely to zero. This, combined with the concept of inflation, brings us close to the idea of the creation of the Universe out of nothing at all, a theme we shall return to in chapter 10.

Meanwhile, having pushed back to the Planck time, the earliest era at which our theories can have even the most tenuous validity, it seems appropriate to digress, briefly, on the ultimate future of the Universe.

The Long-range Forecast

The fate of the entire Universe may well depend upon the nature of one of its smallest constituents. If there are enough dark matter particles around, each with enough mass, then gravity may be sufficient not only to flatten the Universe but to close it, like a black hole, ensuring that it will one day recollapse. The ultimate fate can be determined, in principle, by measuring the way in which the expansion of the Universe is slowing down today—by comparing the recession velocities of objects at high redshift with those of objects closer to home. In practice, this is impossible. We have no reliable way of estimating distances independent of the redshift; galaxies seen at large distances are systematically younger than nearby ones, and so the two cannot be compared directly, unless we know how the intrinsic properties of galaxies change as they age.

For reasons already discussed, in chapter 1, we suspect that there is enough dark matter to decelerate the Universe at just the critical rate associated with flatness. It sits just on the boundary between becoming a black hole and recollapsing, or expanding forever at a measurable rate. If the Universe is flat, it will expand forever, but ever more slowly, so that the recession speed of any galaxy, measured from any other galaxy, will become smaller and smaller. What does the long-range future then hold in store? If the Universe expands indefinitely, there will be enough time for all stars, in all galaxies, to attain a terminal equilibrium. Various timescales are shown in figure 3.11.

10^{14}yr	Ordinary stellar activity completed
10^{17}yr	Significant dynamical relaxation in galaxies
10^{20}yr	Gravitational radiation effects in galaxies
10^{31}–10^{36}yr	Proton decay
$10^{64}(m/m_*)^3$yr	Quantum evaporation of black holes
10^{1600}yr	White dwarfs → iron*
$10^{10^{26}}$–$10^{10^{76}}$yr	Neutron stars undergo quantum* tunnelling to black holes, which then "quickly" evaporate
	*If proton decay does not occur

Figure 3.11 The far future of an ever-expanding universe.

Even the slowest-burning stars would eventually die. All the gas in the galaxies would be tied up in dead stellar remnants—neutron stars, black holes, and cold white dwarfs—and no new stars would form. Galaxies in groups and clusters would merge (our Milky Way will collide with the Andromeda Galaxy, its nearest large neighbour, within 5 billion years, transforming these two disc systems into a huge amorphous starpile). The sky would darken, not just because clusters of galaxies move farther apart as the expansion continues, but because the internal ecosystem dies. Black holes at the centres of galaxies would swallow more and more surrounding gas and stars. If protons do not live for- ever, then all ordinary stars will eventually decay, leav- ing only black holes. These, too, eventually decay by evaporation. If protons did last forever, then the final death of the Universe would be spun out over a much longer period—the longest timescale in the diagram is so enormous that, if written out in full, it corresponds to a 1 followed by a number of zeroes equal to the number of atoms in the observable Universe.

But suppose, in contrast, that there were enough surplus dark stuff to put the cosmic density just above the critical value for exact flatness. If the density were still very close to critical—if the Universe were very nearly flat—the expansion would not reverse until the events described above had all run their course, and all that would collapse would be a Universe of pure radiation. But if the Universe had the highest density that could just fit in with present observations, about twice the critical value, its expansion would halt after another 20 billion years. The redshifts of distant galaxies would then change to blueshifts, as they began to fall inward, and galaxies would eventually crowd together again. Space is already becoming more and more puckered as isolated regions—dead stars and the nuclei of galaxies—collapse gravitationally; this local puckering of the fabric of space would simply be a precursor of a Universal squeeze into a big crunch that engulfs everything. Some key stages in the countdown are shown in figure 3.12. Galaxies merge; stars move faster, just as atoms of gas move faster when the gas is compressed; stars are eventually destroyed, not by collisions but because the sky, filled with blueshifted radiation, becomes hotter than stellar interiors. The final outcome would be a fireball like the one in which the Universe was born, but rather more lumpy and unsynchronised. The earliest this could happen would be about 50 billion years from now—at least ten times the remaining life of the Sun.

The most detailed scientific discussion of the future of the Universe (escatology) is contained in a 1979 article in the journal *Reviews of Modern Physics*, by

t (yr)	
-10^9	Clusters of galaxies merge
-10^8	Galaxies merge
-10^6	Stars moving relativistically
-10^5	Entire "night sky" hotter than stellar surface
-10^3	Stars destroyed
	Black holes grow catastrophically
-1	Temperature $\geqslant 10^8$ K everywhere

Figure 3.12 The fate of a recollapsing universe: countdown to the big crunch.

Freeman Dyson, entitled "Time Without End: Physics and Biology in an Open Universe." He says little about the recollapsing Universe (the mere idea seems to give him claustrophobia) but looks in detail at the future of an ever-expanding universe. He elaborates the points outlined above, and contemplates the outlook for intelligent life. In this perspective, the few billion years that have led to human life on Earth are just a trivial foretaste of the complexity and variety of organisms that might eventually evolve. But can "life," in some form or another, survive and develop intellectually literally forever, storing or communicating an ever-increasing body of information, on finite energy reserves? Dyson shows that, in principle, this can be done. As the background temperature falls, one must keep cooler, think progressively more slowly, and hibernate for long intervals. But infinite timespan does offer unbounded potential for intelligence.

Will our descendants (actual or metaphorical) need to follow Dyson's conservationist maxims to survive an infinite future? Or will they fry in the big crunch a few tens of billions of years hence? The long-range forecast can be refined by observations—by measuring the amount of dark matter in the Universe, or by measuring the deceleration rate. The second part of our book is concerned largely with just this kind of evidence, although the search is at far too early a stage for any such cosmic conclusions to be drawn. But the fate of the Universe, like its present appearance, was imprinted right at the beginning, in the hot, dense fireball era. And to understand that era, and the nature of the relics it could have left behind, we enter the realm of the particle physicist.

PART TWO

———————————⋆———————————

The Stuff of the Universe

CHAPTER FOUR

---★---

The Particle Zoo

PARTICLES COME in many varieties. As physics developed in the present century, the picture showed that these particles interact with one another through the effects of four forces. But the forces themselves are now seen, in the picture derived from quantum theory, as being "carried" by other particles. Two electrically charged objects, for example, exert a force on each other because photons, the carriers of the electromagnetic force, fly between them. *Everything* can be explained in terms of particles, although there is a distinction between particles that carry forces (called *bosons*) and what we might loosely think of as "material" particles, called *fermions*.

Fermions themselves come in different varieties. The ones that matter most to us are the ones that occur in atoms. Protons and neutrons, the relatively massive particles that form the nuclei of atoms, are members of the family known as baryons. Other baryons can be manufactured under conditions of very high energy in colliding-beam experiments; they were also present in the Big Bang. But left to their own devices these heavier particles "decay" into protons and neutrons, releasing energy as they do so. For our purposes, the term *baryonic matter* means protons and neutrons.

Electrons, which are the very light particles found in

the outer parts of atoms, are members of a different family, called the *leptons*. The family consists of three pairs of particles. The electron itself has two counterparts, the mu and tau particles, each much heavier than an electron; each of these three particles has an associated type of neutrino, a very light particle indeed.

Physicists measure mass (strictly speaking, "rest energy") in terms of a unit known as the electron Volt, or eV. In these units, the mass of an electron is a little over 500,000 eV, while the mass of a proton is close to a billion eV, which is called 1 GeV (the "G" stands for *giga*). Each neutron also has a mass of almost 1 GeV; the two common baryons are each some 2,000 times as massive as a single electron. Neutrinos, on the other hand, may have no mass at all. When physicists first realised the need for neutrinos, to explain how energy was being carried away from certain nuclear reactions, they thought that the neutrino might have precisely zero mass, like the photon. But this has never been proved—the appropriate experiments to measure neutrino mass are very difficult to carry out—and the best anyone can say at present is that each electron neutrino certainly has a mass smaller than about 20 eV— that is, less than 0.004 percent of the mass of an electron. Less can be said about the other types of neutrino.

There is one further complication before we can move on from the world of fermions. Protons and neutrons are themselves composite particles, each one being made up of a triplet of quarks. Quarks come in three paired types (like the electron/neutrino pair), varieties that have been given arbitrary and somewhat whimsical names. Two types of quark, called "up" and "down," are present in protons and neutrons; a third type of quark, dubbed "strange," can be produced in high-energy interactions, was present in the Big Bang, and may contribute to the flattening of the

Universe.* But quarks never appear alone, only in groups of three, making baryons, or in pairs, known as mesons, which transfer forces between baryons.

Physicists have developed a satisfactory description of particle interactions involving just two basic families of particles, three pairs of quarks (which occur in different combinations in baryons and mesons) and three pairs of leptons. The symmetry between the two families helps to convince theorists that they have found a fundamental truth of nature.

The two families of "material" particles are accompanied by four kinds of force. One, called the strong force, holds atomic nuclei together and is a result of interactions at a deeper level involving quarks. Another, responsible for radioactive decay, is called the weak force. Neither manifests itself on a scale bigger than the size of the nucleus of an atom—they are short-range forces. Just two long-range forces show up on a human (and larger) scale: gravity, which affects all particles that have mass, and electromagnetism, which affects only charged particles. A possible "fifth force," which has been the subject of discussion recently both in scientific circles and in the press, is really a modification of gravity.

By taking the temperature of the Big Bang, from the microwave background radiation, and calculating the nuclear reactions that would have occurred in the first few minutes of the life of the Universe, physicists estimate that the baryons could provide no more than about 10 or 20 percent of the matter needed to make the Universe flat. (The mass of all the electrons can be

*There are also three other varieties of quark, making up three pairs in all, which are manufactured in high-energy accelerators but cannot exist in the Universe at large today. To keep things simple, we shall ignore them; if you want to know more, see In Search of the Big Bang. The whimsy extends to the names given to some of the properties of these quarks, analogous to electrical charges, which are described as "colours."

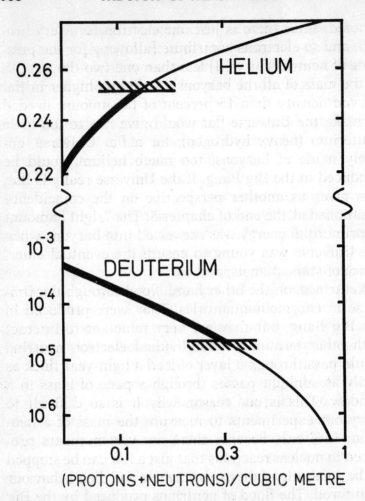

Figure 4.1 The abundances of helium and deuterium emerging from a standard Big Bang are here plotted as a function of the baryon density. (The primordial nuclear reactions proceed faster when the baryon density is high. This slightly raises the resulting helium abundance; deuterium, being an intermediate product on the way towards helium, survives more readily in a *low* density universe.) Helium is made, rather than destroyed, in the course of galactic evolution. The primordial helium therefore cannot exceed the lowest helium abundance measured in old stars, and this sets an upper limit to the baryon density of around 0.1 of the critical density. (The exact value depends on the Hubble parameter *h*.) Deuterium, on the other hand, is destroyed and not made in stars, so its present abundance (about 10^{-5} that of hydrogen) is below that emerging from the Big Bang; primordial deuterium production therefore requires a similarly low baryon density.

ignored, since there is just one electron for every proton, and so electrons contribute [allowing for the presence of neutrons as well] less than one two-thousandth of the mass of all the baryons). A slightly higher initial baryon density than 15 percent of the amount needed to make the Universe flat would give rise to too little deuterium (heavy hydrogen); for a flat Universe, entirely made of baryons, too much helium would be produced in the Big Bang. If the Universe really is flat, this gives us another perspective on the coincidence mentioned at the end of chapter 3. The "right" amount of primordial energy was converted into baryons when the Universe was young to ensure the eventual emergence of stars, planets, and us.

Neutrinos, on the other hand, flood through the Universe in enormous quantities. They were produced in the Big Bang, but they are very reluctant to interact with other fermions. An individual electron neutrino would pass through a layer of lead a light-year thick as easily as sunlight passes through a pane of glass in a window. This is one reason why it is so difficult to carry out experiments to measure the mass of a neutrino; it is only because there are vast numbers produced in nuclear reactions that just a few can be stopped in the apparatus of the experimenters and their behaviour monitored. The flood of neutrinos produced by the Big Bang is a second background in the Universe, like the cosmic microwave background, with 100 million neutrinos left over from the Big Bang in each cubic metre of space even today. If each of those neutrinos had a mass of a mere 30 eV or so, they would contribute enough mass overall to make the Universe flat. Since neutrinos are definitely known to exist, and have been detected in many experiments in laboratories on Earth, as well as being required to balance the books of reactions involving electrons and baryons, they were the prime candidate for the gravitationally dominant stuff of the Universe as soon as astronomers realised the need for dark matter. Unfortunately, they may not quite fit the bill.

Sizing Up Neutrinos

The first firm evidence that neutrinos exist came only in 1956, and the idea that they were massless held sway for almost a quarter of a century after that. In 1980, however, Valentin Lubimov and colleagues at the Institute for Theoretical and Experimental Physics (ITEP) in Moscow claimed that they had measured a mass for the electron neutrino, and quoted a figure of between 14 and 46 eV. This was enough to cause a flurry of excitement among astronomers, since the range of masses covered by the claim was just right for neutrinos to provide all the matter needed to flatten the Universe. As a result, a great deal of work was carried out on hot dark matter models of galaxy formation, but by the end of the 1980s it was becoming clear that these models, in which *all* the dark matter is in the form of neutrinos or other hot particles, cannot easily be made to fit the observed distribution of galaxies across the Universe.

Meanwhile, other experimenters had been trying to verify the Soviet claims. The experiments are horribly difficult, which is why the masses "measured" are usually quoted as a broad spread of possible values, or as an "upper limit." It is easier to prove that the mass of a neutrino must be less than a certain amount than it is to give a precise value for the mass. A Swiss team, carrying out the same kind of experiment as the ITEP group, found an upper limit of 18 eV, while a team at the Los Alamos National Laboratory in New Mexico said that the mass must be less than 27 eV. All of those claims can just about be reconciled with one another.

But in 1987 the latest results from ITEP quoted a "best value" of 30.3 eV, and a possible range of electron neutrino mass between 22 eV and 32 eV. The Soviet data seem to be out of step with other studies, a pattern confirmed by analysis of neutrinos that reached the Earth from a supernova that exploded early in 1987.

A supernova occurs when a certain kind of large star at the end of its life (which may be quite short compared with the life of the Sun) runs out of nuclear fuel to keep itself hot. The core of the star collapses, converting a great deal of gravitational energy into heat. The energy released explodes outwards, blowing apart the outer layers of the star. So much energy is released that for a few weeks a single supernova star can shine as brightly as all the ordinary stars of the Milky Way put together. Such events are rare, and when a star blew up in the Large Magellanic Cloud, a neighbour to our Milky Way, it was the closest supernova seen since the invention of the astronomical telescope. At that, it was about 170,000 light-years away; the light that astronomers saw from the explosion had left the Large Magellanic Cloud, by terrestrial reckoning, during the ice age before last.

Studies of observations of this rare event are providing new insights into how stars work, and have confirmed that all elements heavier than helium are manufactured inside stars and are spread through the cosmos by such explosions. The supernova has also provided a new estimate of the mass of the electron neutrino.

Several experiments being run at different sites on Earth are capable of detecting sufficiently energetic neutrinos (and, of course, antineutrinos; we use the term *neutrinos*, without qualification, as shorthand for both). These detectors are designed primarily for other work, monitoring other forms of radiation, but when an antineutrino interacts with the nucleus of an atom in the detector it produces a positron with characteristic properties, and such events can usually be identified. At a distance of nearly 170,000 light-years, 22 neutrinos shared among three detectors (in Japan, 11; the United States, 8; and the Soviet Union, 3) were unambiguously recorded from the supernova; 5 other neutrinos arrived at a fourth detector, under Mont Blanc, five hours before, but this may have been a coincidence.

That doesn't sound like a lot, and since neutrinos are so notoriously reluctant to interact with other forms of matter, it may seem a surprise that any neutrinos from supernova 1987A, all that distance away, were detected on Earth. So, how many were produced by the event?

The simplest way to estimate the production of neutrinos in the supernova is to note that the collapsing core contained about as much mass as our Sun, about 2×10^{57} baryons every proton in the core was converted into a neutron, with a neutrino being released as a result. The collapsing and bouncing core gets hot enough to emit several times more neutrons in other ways, making a total of around 10^{58}. A 1 followed by 58 zeroes is a big number, but is it big enough to provide more than a few neutrinos at a distance of 170,000 light-years? The way to find out is to work out the surface area of a sphere 170,000 light-years in radius, and divide that into the number of neutrinos available.

Some idea of just how big a number 10^{58} really is can be gleaned from the calculation, which shows that *a hundred billion* (10^{11}) neutrinos from the supernova passed through every square centimetre of the Earth's surface in a few seconds during February 1987. About 10^{14} neutrinos passed through every person on the planet, including you—and you didn't feel a thing. In all that flood, on average one person in a thousand had a single supernova neutrino stopped by their body; one person in a hundred million had a supernova neutrino interact inside his or her eye. There actually may be people around who "saw" a flash of light caused by the explosion of a star in the Large Magellanic Cloud, if the interaction happened to trigger nerve endings in the eye. The fact that only 22 neutrinos out of all that flood stopped in detectors on Earth shows just how reluctant neutrinos are to be stopped by anything.

Presumably, all those neutrinos were produced in the space of a few seconds, as the core of the star collapsed. But the little pulse of neutrinos that stopped in the detectors was spread over about a dozen seconds. They could have got spread out a little on their journey, but not by much, and this provides a clue to their mass.

If neutrinos had precisely zero mass, like photons, then like photons they would travel at precisely the speed of light, and they would all arrive together if they set out together. If each neutrino has a tiny mass, however, they will not all move at the same speed. Each will move very nearly at the speed of light, but neutrinos with more energy will travel a tiny bit faster than their companions and will arrive first. Of course, it may be that the neutrinos diffused out of the supernova in a burst spread over twelve seconds, and that they have all travelled at the speed of light. But *if* they were produced together and the spread is due to the mass effect, the size of the spread (twelve seconds after a journey of 170,000 light-years) provides a measure of how small the neutrino mass must be.

There are several ways to tackle the calculation; by taking all 25 neutrinos together, or by using the data from the three different detectors separately. The most clear-cut interpretation of the evidence is that the mass of each neutrino must be less than 20 eV, which fits in with all the experiments on Earth except that from ITEP. One set of calculations is consistent with the possibility that the neutrino has a definite mass of about 3 eV.

There is scope here for endless debate, which will be settled only when another supernova explodes nearby, or when the Earth-based experiments become more accurate. But there is already enough information to help us in our quest for the stuff of the Universe. A mass of less than 15 eV is certainly not enough for electron neutrinos to flatten the Universe unaided, without fiddling the cosmological calculations uncomfortably. It is still possible that neutrinos have a tiny mass, and they could play a part in determining the dynamics of galaxies and in *helping* to flatten the Universe. But other relics of the Big Bang, in a form never yet detected here on Earth, could be more important. We do not know enough about such particles to predict their masses, nor how many should survive from the early Universe. Nevertheless, it may well be that 90 percent of the Universe is in this unknown form. Possible candidates proliferate in the particle zoo.

Missing Links

It is intriguing that just at the time, during the 1980s, when astronomers were beginning to worry about the need for dark matter to flatten the Universe, particle physicists, for quite different reasons, were finding a need for undetected forms of dark matter ("new" particles) to complete their theories. These are well-founded theories, backed up by experiments, which are a step on the road to a single unified theory of all the forces of nature. They explain very well our observations of the world of the very small—but *only* if there are types of particles around that have not yet been detected directly. The fact that, by and large, the requirements of the astronomers are met by the kinds of particles invoked by the particle physicists to complete their theories is surely a sign that both are on the track of the truth. From opposite ends of science, the very large scale and the very small scale, the signposts on the path of future research point the same way.

The key concept in the successful theories of particle physics today is symmetry. This concept crops up at several different levels. There is a symmetry, for example, between particles and antiparticles—a positron is the opposite of an electron in every possible way, and one manifestation of this is that it carries positive charge instead of negative charge. Most interactions involving elementary particles work just as well if we imagine reversing the direction of time, so that the reactions run backwards. In fact, some interactions are not time-reversible, but these rare exceptions to the rules of symmetry provide insights that have helped in the development of the standard model of particle physics.

Another kind of symmetry can best be understood in terms of energy. In the everyday world, the energy needed to raise an object through a small distance (say, one metre) can be measured. It doesn't matter whether we start the experiment with the object on the floor and raise it onto a desk top, or whether it starts on the desk and ends up near the ceiling. It doesn't matter whether we carry out the measurement on one side of the room or the other. The amount of energy involved is always the same; what matters is the *difference* in height between the starting point and the finishing point, and in principle (ignoring such inconveniences as friction) the energy required is the same whatever route the object has followed between its starting point and its finishing point. This is a manifestation of a property of the world known as *gauge symmetry*.

Gauge symmetry is an inbuilt feature of the standard model of particle physics. It has helped physicists to combine their descriptions of the electromagnetic force and the weak force into one mathematical package, the electroweak theory. And it has raised hopes of combining the nuclear force as well into one package. But the successes of these theories depend on building in more symmetry than we actually see in the particle world. These symmetries would exist at very high energies (for example, in the Big Bang), but are hidden in the everyday world. Breaking, or concealing, such a symmetry requires the presence of extra particles in the Universe. At the same time, there are features of the standard model that look as if they ought to be symmetric, but in which the symmetry can be preserved only if there are particles around that have yet to be discovered.* Either way, some of these particles, the missing links in the standard model, are candidates for the dark matter.

The Axion

The missing link that seems to have the best status in terms of particle physics today is called the *axion*. The need for such a particle arose in a straightforward fashion from studies of some of the symmetries of particle physics, and it also fits the needs of the cosmologists.

If every particle involved in an interaction were replaced by its antiparticle counterpart, the interaction would still proceed. Swap every electron for a positron, every proton for an antiproton, and so on, and physics still works. This symmetry is known as *charge conjugation*, or C. The symmetry that says that an interaction will proceed just the same if everything is swapped left to right, as if reflected in a mirror, is known as *parity*,

---

*Although physicists didn't think in those terms at the time, the discovery of the positron, an antimatter counterpart to the electron, hinted at a previously unsuspected symmetry and showed that there must also be an antiproton and an antineutron, and all the other antiparticles, in order to preserve symmetry. Present-day reasoning follows similar lines to conclude that there must be undetected particles in the Universe.

or P, and the symmetry that says that interactions can equally well proceed forwards or backwards in time is given the label T.

These symmetries are true for most interactions but do not always hold perfectly on their own. Interactions that involve the strong force, however, turn out to have the property that they are *always* symmetric when both these changes are carried out together—the interactions show a combined CP symmetry, in such a way that if C and P are both violated, the differences cancel out in the CP symmetry. As they originally stood, the equations that describe the strong force should have had this symmetry built into them. But Gerard t'Hooft, a Dutch physicist, found that under some circumstances the equations allowed both CP and T to be violated. It was only in 1977 that Roberto Peccei and Helen Quinn, at Stanford, found that the required symmetry, preventing such violations, could be built back into the equations by adding in the effects of a new type of particle, the axion. Since CP and T are never found to be violated for the strong force, this is a powerful argument that such a variety of particle must exist.

No experiment has ever detected an axion. This is not surprising, since the theory says that axions should be as elusive as neutrinos. However, the theory also says that axions could be very much more common than neutrinos, and that they must have some mass, however small. If the theory is correct, axions form a third background left over from the Big Bang. If each axion has a mass of one hundred-thousandth of an electron Volt (10^{-5} eV), and they are present in the quantities required by the theory, they would flatten the Universe. In spite of their tiny mass, however, axions are born in the Big Bang with very small velocities; they are good candidates for the cold dark matter required by many successful models of galaxy formation.

Supersymmetric Partners

The ultimate development of the idea of symmetry is a family of theories that together are known as super-symmetry. On this picture, symmetry extends to the point where every type of boson must have a fermionic partner, and every type of fermion must have its counterpart in the world of bosons. These theories are attractive because they seem to succeed in bringing gravity into the overall picture and unifying it with the other forces of nature in one mathematical description. Such theories are, as yet, incomplete, but promising. The obvious drawback to them, at first sight, is that although each fermion type and each boson type is supposed to have a counterpart, none of the known bosons fit the bill as partners for the known fermions, and vice versa.

The only way out of this difficulty is to suppose that nature does actually provide scope for partners to every known type of particle, but that these partners have never been detected. At a stroke, supersymmetry doubles the number of inhabitants of the particle zoo. Hypothetical partners to fermions, in this scheme of things, are given names beginning with *s*, so that, for example, it is assumed that the electron is matched in the boson world by the selectron. Similarly, the hypothetical fermionic counterparts to bosons are given names ending in *ino*, and the photon, for example, is regarded as having a counterpart known as the photino. So where are all these extra particles?

Even in the familiar world of everyday fermions, most of the known particles are unstable. Heavy particles, created in laboratories on Earth or left over from the Big Bang, give up energy and decay into familiar protons, neutrons, and electrons plus neutrinos. Indeed, even a neutron isolated outside an atomic nucleus will decay within a few minutes, spitting out an electron and (anti)neutrino and becoming a proton. Fundamentally, only the proton and electron, and neutrinos, are stable.* The supersymmetric partners of the familiar particles would behave in much the same way, and only the lightest of these particles (the lightest supersymmetric partner, or LSP) is thought to be stable and likely to be around in any large quantities in the Universe today. In different versions of supersymmetry, different particles have the honour of being the LSP; the mass of the LSP is not well pinned down, but could be anywhere from a few times the mass of a proton—a few GeV—to one hundred times the mass of the proton. The best candidates today are the photinos, electrically neutral particles that are the fermionic partners of photons. If the photino is the LSP, it lives forever.

Photinos, gravitinos (supersymmetric counterparts to gravitons), and axions may seem like exotic inhabitants of the particle zoo, but although they have not been detected yet, there are sound theoretical reasons for expecting them to exist, and some of our cherished theories would fall apart without them (it is, of course, possible to have both axions *and* supersymmetry). Those theories are not cherished without good reason; they work very well as a description of the particle world, and if they fall apart, then most fundamental physics research since about 1950 falls apart with them. Some other candidates for the role of dark matter are not quite in that league. Some theories *allow* these even more exotic particles to exist, but they are not *essential;*

---

*Even protons may decay on a timescale of 10^{33} years or more, as discussed in *The Omega Point*. This is not directly relevant to our present story, but it is worth a passing mention, since the detectors that spotted neutrinos from the supernova in 1987 were actually designed to detect the products of proton decay, on the principle that if one proton decays in 10^{33} years, a tank of water containing 10^{33} protons, in the form of hydrogen atoms, ought to produce one decay each year. So far, no neutrinos produced by the decay of a proton have yet been seen.

nobody is going to be too upset if they turn out not to be present in the Universe, while most physicists would probably be relieved to do without them. Even so, a couple of at least semiserious prospects deserve a mention.

Making the Most of Monopoles

Electrical charge comes in two varieties, positive and negative, and it is quite possible to have an isolated charge of either kind—the electron, which carries negative charge, is an example. Magnetism, which is so much like electricity that they are jointly described by the same set of equations, as electromagnetism, also comes in two varieties, north and south; but an isolated magnetic pole is never seen. Try to cut a bar magnet in half, to separate the north and south poles, and you will be left with two smaller, but perfectly formed, bar magnets, each with its own north pole and south pole. This contrast with the behaviour of electrical charges has provided mathematical physicists with something to puzzle over for more than half a century.

In 1931, the pioneering quantum theorist Paul Dirac showed that in ordinary quantum theory magnetic monopoles are, in fact, allowed to exist. They are not obligatory, but there is nothing in the equations that forbids them, and it is a rule of thumb in physics (not intended entirely facetiously) that anything that is not forbidden does happen, somewhere and sometime. But where, and when, do monopoles form?

In 1974, different teams of researchers investigating the nature of unified theories, the foundation stones of the standard model of particle physics, both came up with the same discovery. In the favoured forms of those unified theories, symmetry breaking as the Universe cools from the high-energy state of the Big Bang to its present state *must* result in the production of free magnetic monopoles. In those theories, monopoles are not optional but obligatory. What's more, the theories tell us how much mass each monopole must have—10^{16} times the mass of a proton.

There is no way that such a massive particle could be created in an experiment on Earth today. Such experiments "make" particles by putting enough energy into the colliding beams of accelerators like those at CERN so that $E = mc^2$ can convert the energy into massive particles, which, hopefully, can be tracked for a while before they decay into other particles. Each run of those machines is a direct confirmation of the accuracy of relativity theory. The decay would not be a problem for monopole seekers, since monopoles ought to be stable; but the mc^2 *is* a problem. Although this is only the equivalent of the energy released when a hand grenade explodes, concentrating that energy into a pair of colliding particles at the subatomic level requires a collision 10^{14} times more energetic than can be engineered in any particle accelerator on Earth.* There was, however, ample energy available in the Big Bang to make anything at all (if you go back far enough towards the moment of creation), and so there is some theoretical basis for expecting a fourth background, of magnetic monopoles, to have emerged from the Big Bang.

With so much mass in each one, it wouldn't take an impossible number of monopoles to flatten the Universe. But no magnetic monopole has ever been detected on Earth, and no effects caused by magnetic

---

*Even with unlimited energy, you might not be able to make monopoles in a simple colliding-beam experiment. The theories that make monopoles obligatory regard these "particles" as defects in the structure of spacetime, cracks in space like the cracks that appear in an ice cube when it is dropped into a drink. To make these defects, you need to start from a high-energy state of the whole Universe.

monopoles have ever been seen in the Universe at large. We certainly could detect them, if they are there—if there are enough monopoles around to flatten the Universe then thirty of them should pass through any area on Earth the size of a soccer field each year; they would also tend to neutralize or short out the magnetic fields that pervade our whole Galaxy.

It is something of a theoretical puzzle why monopoles *haven't* been detected, if they can be created in the Big Bang. The best resolution of this puzzle is provided by the various forms of the theory of inflation, which says that at a very early time a tiny piece of spacetime expanded dramatically to form the Universe as we know it. (This inflation, remember, is postulated in order to explain the smoothness and uniformity of our Universe.) It turns out that the region of spacetime that expanded so dramatically during inflation would have been so small that it could have contained only one or two monopoles. Unified theory could be correct, and magnetic monopoles may indeed exist, but with only a couple of them around in the whole visible Universe it is unlikely that we would ever be aware of their presence. It is still *possible* that monopoles abound in the Universe, but even monopole enthusiasts admit that this is unlikely.

Quark Nuggets

Another candidate, which is equally implausible but deserves to exist as a reward to the ingenuity of the theorist who dreamed it up, is the quark nugget. The idea came from Ed Witten, of Princeton University, in 1984; once again, it derives from the highly energetic conditions that prevailed during the Big Bang.

Ordinary baryons are made only of up and down quarks. Part of the reason for this is that strange quarks are much more massive and cannot easily be formed by natural processes today. During the Big Bang, however, there was ample energy around, and some of this E could be converted into the mc^2 of strange (or s) quarks. Witten suggested that a "quark nugget" made of roughly equal numbers of u, d, and s quarks could be stable— more stable, in fact, than ordinary u-d matter. He showed that the kind of quark nuggets that might have emerged from the Big Bang would each have a radius in the range from 0.001 centimetre to 10 centimetres, with corresponding masses in the range from 10^6 to 10^{18} grams (no messing about with electron Volts here!); they would be denser than the matter inside a neutron star. If enough matter to flatten the Universe existed in the form of such nuggets, spread through the Universe, an average of 10^9 grams ought to be swept up by the Earth each year as it moves through space. This corresponds to a range of possibilities from one big nugget hitting the Earth once in its lifetime, to a thousand little nuggets showering down on us each year.

There is, of course, no evidence that any quark nugget has ever hit the Earth, and no evidence that such objects exist at all. They are not a very likely candidate for the dark matter, which is unfortunate because the nugget concept leads naturally to a universe in which the respective contributions of ordinary and dark matter do not differ by more than about a factor of 10. The coincidence that there is a fairly even split between the mass in ordinary baryonic bright matter and in nuggets falls naturally out of the model. Alas, Witten's calculations suggest that the energetic neutrinos in the early Universe would destroy quark nuggets, so they would not survive unless they were as massive as a planet, about 10^{27} grams; it is very difficult to see how nuggets containing as much mass as a planet, or more, could form in the first place. We won't yet scratch quark nuggets from our list of possibles, but at present they look like very unpromising candidates for the dark stuff.

A Black Hole Bonanza

Much the same can be said about mini black holes. Black holes have become part of popular mythology, entering the realms of scientific folklore, and many people who know little of science have heard of these objects. In case there are any misconceptions about them, however, perhaps we should sketch out their general features before concentrating on the kind that could, conceivably, provide part of the dark matter.

The kind of black holes that occasionally make headlines in the nonscientific press is associated with the collapse and death of stars. The force of gravity holds things together and pulls all the particles that make up a star towards its centre. As long as the star contains enough nuclear fuel, it can keep itself hot in the centre by converting lighter elements into heavier elements. At each step in this process, a little mass is converted into energy. The resulting heat provides a pressure that holds the star up against the tug of gravity. But when all the nuclear fuel is exhausted (which means when the core of the star is made of iron, the most stable element), no more energy can be produced in this way.

One possibility is that the star will then collapse so suddenly, releasing gravitational energy, that a supernova, like 1987A, is born. This is probably inevitable for any star with a mass, at the end of its life, more than eight times bigger than the mass of our Sun. Smaller, and more common, stars simply cool and contract slowly once the internal source of heat is exhausted. But how far will such a star collapse?

Provided the star ends up with the mass of our Sun, or less, there is no problem. The baryons inside the star pack tightly together as it cools, and eventually it becomes a dwarf star, containing as much stuff as the Sun in a sphere the size of the Earth—essentially a giant crystal of iron. At this point, it is held up by quantum forces, which do not permit the particles to squeeze more closely together.

If the dead star has a little more mass, however, then gravity can overcome these quantum forces. Electrons are forced to combine with protons to make neutrons. The whole star shrinks in upon itself as a result, becoming a ball of neutrons, a neutron star. It may have a little more mass than our Sun, but it occupies a sphere no bigger than a terrestrial mountain. It is, in effect, a single "atomic" nucleus. Once again, quantum forces are holding it up against the inward tug of gravity.

With a little more mass still, not even quantum forces can stop the collapse. The fate of any cold object containing more than a few times the mass of our Sun is to collapse completely, into a mathematical point, a singularity. As it does so, it wraps spacetime around itself, forming a black hole.

Gravity distorts spacetime, and gravity is stronger in the immediate vicinity of denser objects, so spacetime is increasingly distorted as an object with a particular mass shrinks. At a critical size, known as the Schwarzschild radius, spacetime bends completely around on itself, cutting the region of spacetime inside the hole off from the rest of the Universe. The Schwarzschild radius is smaller for smaller masses and larger for more massive objects; the Schwarzschild radius corresponding to the Sun's mass is 3 kilometres, while for the Earth it would be a bit less than a centimetre. Things can still fall into the black hole, but a sphere with the Schwarzschild radius marks a boundary, a surface around the hole, from which nothing, not even light, can escape. This "Schwarzschild surface" is in effect the surface of the black hole. At the singularity inside, the laws of physics break down, and we can have no knowledge of what happens then. One bonus of the formation of the black hole is that it screens us from the singularity, and stops us from being confronted with a point where the laws of physics would be transcended in some as-yet-unknown way.*

This scenario applies to black holes that contain about the same amount of stuff as the Sun—stellar mass black holes. Much bigger black holes may exist in the Universe, supermassive black holes each with the mass of a hundred million Suns, lurking in the hearts of galaxies and chewing up any matter that comes their way, including not just gas but entire stars and even whole clusters of stars. Although nothing can escape from inside the Schwarzschild surface, in the swirling maelstrom of material funnelling into such a black hole so much energy is released that the object may be seen at a redshift of 4, as a quasar.

In order to close spacetime around a small object, a star the size of a mountain, the spacetime has to be bent very tightly, by a strong gravitational field. A much larger region of spacetime could be closed off from the rest of the Universe by a much more gradual bending. Athletes running on an indoor track usually have to follow much tighter bends than when they run outdoors; in both cases they run around the track and

---

*We run into a similar "singularity" as we extrapolate the Big Bang back towards "time zero." But a black hole singularity exists only in the future, not the past, for any black hole that forms in the contemporary Universe. The very existence of our Universe depends on the unknown physics at the cosmic singularity. However, the suspected singularities in black holes cannot affect us; their existence does not diminish our confidence in predicting what happens *around* black holes, any more than chemistry (which deals largely with the behaviour of electrons in the outer parts of atoms) is rendered unusable by our lack of detailed knowledge of what goes on inside the nucleus of an atom.

back to where they started, following closed paths, but the outdoor track is bigger and its curves are more gentle. All aggregations of matter do bend spacetime, and even a rarefied object could, in principle, "close up" space if it were large enough. The popular image of the black hole at the heart of a quasar is that it is a region of highly distorted space, where gravity crushes matter out of existence—but such a black hole can actually be formed from a few tens of millions of solar masses of material at a *density* no greater than that of water here on Earth. The "superdensity" at the heart of a quasar is actually matter packed no more tightly together than the atoms of water in the Atlantic Ocean— for the largest holes whose effects we observe, the density is more than the air we breathe. The black hole effect becomes more prosaic still as we look to bigger and bigger scales. Indeed, if the curvature of the Universe were sufficient to make it "closed" (that is, with a density slightly greater than that required for "flatness"), then its eventual collapse would resemble the formation of a black hole—except that we would be inside it! It would be a nice irony if the main concentration of matter that made spacetime curve in this way were itself locked up in much smaller black holes scattered across the Universe.

Stellar mass black holes, or even the kind that power quasars, count as baryonic matter for the purposes of describing the stuff of the Universe. There is no way we can put a label on the kind of matter that is inside such a hole today; what matters is that these holes were made after the Big Bang, so that the matter that went into them was the kind of matter we see in the Universe today. A stellar mass black hole is made of the stuff of a star, which means baryons; a supermassive black hole would also have formed by accreting mainly baryons, in the form of stars and gas. But there is another kind of black hole.

In principle, any amount of matter could be squeezed into a small enough volume of space to produce such an intense gravitational field that it would close space-time around itself and become a black hole. In fact, no conceivable process operating in the Universe today could "implode" an object much less massive than our Sun by the required amount. The *less* mass the object has, the *harder* it is to make a black hole—with enough mass, the hole simply makes itself. As always, however, you can invoke enough energy to do anything at all if you look back far enough into the Big Bang towards the moment of creation.

"Miniholes" could exist as fossils of fluctuations in a very early era when the whole Universe was more dense than the nucleus of an atom, or the inside of a neutron star, is today. Because they would have formed when the energy in the fireball was not in the form of baryons but in ultrahot radiation, they do not count as baryonic matter, and could provide the required mass to make the Universe closed. At one time, it looked as if they might even be detected in the Universe today; then, the failure to detect traces of miniholes seemed to rule them out as dark-matter candidates. Now it seems that they might be lurking in the Universe undetected after all.

Do Black Holes Explode?

Stephen Hawking, of Cambridge University, investigated the behaviour of miniholes in a series of calculations he carried out in the 1970s. He showed, in a famous phrase, that small "black holes aren't black," and that they can radiate away energy, shrinking as they do so. The Hawking effect is intriguing because it links the three basic concepts of physics. General relativity comes in, because gravity is involved; thermodynamics is there, because each black hole has, according to Hawking, a specific temperature; and quantum mechanics provides the mechanism by which the holes evaporate.

The simplest way to think of this is in terms of "virtual pairs" of particles created at the edge of the black hole. Far away from any such hole, in the vacuum of ordinary flat spacetime, such pairs of particles are constantly popping in and out of existence. An electron and a positron, say, may pop out of nothing at all and almost instantly disappear again (it has to be a matter-antimatter pair, for symmetry reasons). This strange dance of virtual particles is a requirement of quantum mechanics, linked with the concept of uncertainty, which says, in effect, that anything can happen, provided it happens quickly enough. It sounds like Alice in Wonderland, but in fact the existence of virtual particles helps to give the vacuum its properties and explains details of the way the electrical force, for example, operates—details that cannot otherwise be accounted for.*

In flat spacetime, virtual pairs appear and are gone again far quicker than the blink of an eye, in about 10^{-20} seconds. Next to a black hole, however, things are different. There, in the split second that a pair of particles exists, one may be captured by the intense gravitational field of the hole, falling inwards, never to be seen again. The second particle in the pair scoots outwards from the hole. The mass-energy of the new particle comes from the mass-energy of the hole itself, gravitational energy converted into mass in the region of highly distorted spacetime. Particles boil off from the surface of a black hole all the time, while the hole itself shrinks.

For a large hole, this effect is of no significance because it gobbles up matter far more quickly than it

*See *In Search of Schrödinger's Cat.*

loses mass. A tiny black hole, however, shrinks as it evaporates. A runaway process ensues, with the last dregs of energy being released in a burst of intense radiation, as gamma rays. Quantum effects involving gravity may halt this process, stabilising the black hole at the point where the two "forbidden zones" of figure 3.1 (p. 64) meet. This would avoid the embarrassment of having the central singularity exposed to view, but the mass of the remaining hole would only be about 10^{-5} grams.

The smaller the hole is to start with, the sooner this explosion will occur. Miniholes that each have a mass of about 10^{15} grams (the mass of a cubic kilometre of rock, but the size of a single proton), created in the Big Bang, should be exploding in this way today, 15 billion years after the Big Bang. The observed background of gamma ray energy coming from the sky is so low, however, that it seems that very few such holes are now evaporating, and that comparably few lighter holes have exploded during the history of the Universe to date.

The nature of the final explosion depends on details of particle physics at high energies—in particular, on how many different species of particle exist. If this number were large, the final explosion would create an expanding fireball where many of these particles would decay into electrons and positrons. This fireball would sweep up the magnetic field in the surrounding space, converting its energy into a radio pulse. Radio telescopes are so much more sensitive than gamma ray detectors that they could record such an event—the explosion of an entity smaller than a single proton—from as far away as the Andromeda Galaxy. We could then discover miniholes even if they were far too rare to contribute significantly to the cosmic mass density, or even to the gamma ray background.

Holes with rather lower initial masses, down to about

10^9 grams, might also be ruled out by other considerations. Such lower-mass holes would be hotter (more energetic) and would create more massive particles. Whereas a hole with a mass around 10^{15} grams would probably manufacture only electrons, the lower-mass holes might make baryons, producing changes in the amount of helium and deuterium in the Universe that might be detectable, from studies of old stars, but are not in fact seen.

That left the possibility that more massive black holes could contribute to the dark matter. They are an implausible candidate, because the initial conditions would have to be set just right, "fine-tuned" to produce such large miniholes without making smaller ones as well. In addition, the structure of such a universe would have to be rough enough for these holes to form, without being too rough on larger scales, where we know, from observations, that the Universe is smooth. There is no firm reason to expect this, and therefore miniholes are "also rans," like quark nuggets.

Great Dark Hopes

There is no shortage of cold dark matter candidates. It is worth stressing, however, that the conjectures we have described are not simply wild flights of fancy but are well-motivated ideas. The motivation comes from the success of models in which the Universe is dominated gravitationally by cold dark matter, successes that include correct predictions of many properties of galaxies, including their masses and the way they are distributed in space. These successful predictions apply irrespective of the identity of the dominant CDM particle; it could be any of the ones we have mentioned, or two or more of them present in significant quantities in the Universe, or something else entirely if we haven't yet hit on the correct solution to the puzzle.

Of all the candidates assumed today, axions or photinos are the most promising. The axion fits the requirements of astronomy, and it also solves the only real problem with the standard model of particle physics, which is preserving the symmetry of interactions involving the strong force. The photino, or some other supersymmetric particle (generically referred to as "inos"), is an equally attractive idea. Several groups of physicists around the world are now designing and building experiments to detect such particles (see chapter 5).

If neither the axion nor the photino stands up to further inspection, then it might prove easier for the astronomers to modify their theories to accommodate hot particles, in the form of neutrinos with a mass of about 20 eV, than for particle physicists to tailor their requirements to match astronomical observations. Never forget one great merit of the neutrino—it is actually known to exist! Even better, with neutrinos you get three tries for the prize, since neutrinos come in three varieties. Although electron neutrinos with mass as high as 20 eV now seem unlikely, mu or tau neutrinos have not yet been excluded.

The other candidates really are dark horses. Monopoles, quark nuggets, and miniholes are delightful ideas. If we are being scrupulously honest, however, and not deluding ourselves with wishful thinking, we have to admit that they are rank outsiders.

Which leaves us with one form of stuff that has not received much attention in this chapter—baryons. According to the standard model of the Big Bang, baryons fall short of the critical density by a factor of around 10. There could only be enough of them to flatten the Universe if the physical conditions in the first few minutes differed from those of the standard theory, so that helium and deuterium could still emerge in proportions consistent with observations. Nevertheless, luminous stars and galaxies provide only 1 percent of the critical density, so even in the standard model bright stars need not be the only form of baryonic matter (or even the main form of baryonic matter) in the Universe. What about huge black holes, each one containing the mass of a million Suns, which do not happen to lie at the hearts of galaxies and so do not gobble up matter and belch out energy? What of very faint stars, or planetlike objects, that drift through space unobserved simply because they are too dim to be visible from Earth? Before we get too carried away with exotica, perhaps we should take stock of how much dark but baryonic stuff there might be around in a galaxy like our own—and also take a look at where nonbaryonic stuff might also show up on the scale of galaxies and clusters of galaxies.

CHAPTER FIVE

★

Halo Stuff

WE KNOW THAT there is more to a galaxy than meets the eye, because astronomers can study the way galaxies rotate. The invaluable redshift tells us how fast the stars and gas are moving in different parts of a disc galaxy. You might expect the motion of stars around the centre of a galaxy to resemble the way planets move around the Sun, with more distant planets (stars) moving slower and taking much longer to complete their orbits. That, indeed, is what astronomers expected to find when they began making these measurements. But they were surprised.

In our Solar System, the closest planet to the Sun, Mercury, zips around in its orbit once every 88 of our days. The year here on Earth is just over 365 days long, and the distant giant planets take tens of our years to orbit the Sun once. This is quite different from, say, the way a record rotates on the turntable of a hi-fi system. There, the whole disc rotates as a solid object, and the time it takes for a point on the outer edge to complete one revolution is exactly the same as the time it takes for any other point on the disc to complete one revolution. In the Solar System, the outer planets not only take longer in their orbits but actually move more slowly through space than do inner planets, with or-

132

bital speed decreasing as the square root of their distance from the Sun. In a disc galaxy, the stars farther out from the centre do take longer to complete their orbits, but there is an important difference from the way the planets orbit the Sun. In a galaxy, the velocity of a star is *not* less if it is farther out from the centre. Stars sweep around at the same velocity regardless of their position, and in particular regardless of their distance from the central nucleus of the galaxy. It takes longer to complete the orbits farther out simply because the orbits are bigger. Furthermore, this "law" even affects stars and gas farther beyond the limit of the bright luminous disc. The outer gas, studied by radio astronomy techniques, is "feeling" more matter than the gas farther in. This extra matter must be dark. That evidence shows that there may be as much as ten times more dark matter than bright stars even in a galaxy like our own. The way this dark matter interacts with the bright stuff of a galaxy provides an example of another cosmic coincidence, or a kind of cosmic conspiracy. If a galaxy were made of the bright stuff we see

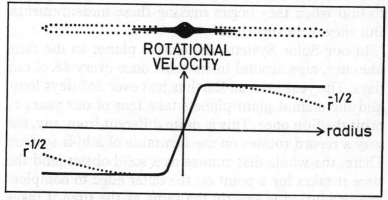

Figure 5.1 A schematic picture of the rotation curve for a disc galaxy. The curve remains flat even way out beyond the apparent edge of the bright disc: the rotation speed does not fall off as $r^{-1/2}$, as in Kepler's law. The gas and stars further out are "feeling" an extra gravitational pull, due to a halo whose density falls off as r^{-2} (so that the mass within radius r increases in proportion to r).

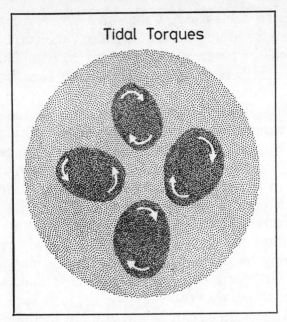

Tidal Torques

Figure 5.2 Protogalaxies would not be spherical; their mutual tidal interactions would therefore set them slowly rotating.

alone, the outer parts would rotate more slowly than inner regions. The presence of a dark halo could change this pattern in almost any conceivable way, depending on the amount of dark matter, its distribution, and the way it moves. In fact, in every case the dark stuff boosts the rotation of the outer parts of the bright galaxy just enough to ensure that all parts of the bright galaxy rotate at the same speed. Such an unexpected coincidence must be telling us something fundamental.

The insight actually seems to tell us how galaxies were made. Any theory of galaxy formation must explain why disc galaxies are rotating. A rotating object spins faster when it shrinks inward, so the rotation (angular momentum) cannot have been "stored" in the Big Bang. The spins were probably acquired through tidal interactions between neighbouring protogalaxies

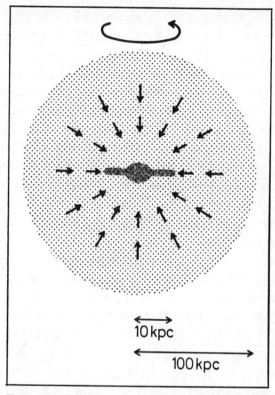

Figure 5.3 Tidal torques give a protogalaxy only about a tenth of the spin needed for rotational support. The gas that forms a rotating disc galaxy with radius 10 kpc must therefore have been "spun up" by falling in from about 100 kpc. If the gas contributed only 10 percent of the protogalaxy's total mass, the rest being a nonbaryonic dark halo whose density varies as r^{-2}, then the disc and halo would contribute equally to the mass at 10 kpc. We can thus understand why the inner (disc-dominated) and outer (halo-dominated) parts of the rotation curve (figure 5.1) join smoothly.

before they collapsed—the mutual gravitational tug of two galaxies would give them equal and opposite spins. But the spins that can be acquired in this way, simple calculations show, are only one-tenth of the amount needed for centrifugal force to balance gravity. If, however, the spinning gas cloud began to settle towards the centre of a gravitational pothole, it would spin up and form a stable, centrifugally supported disc with a tenth of the radius of the initial halo. Because the gas contains just about one-tenth of the total mass (the rest being in the dark halo), the disc ends up with its own gravity, in the inner part of the pothole, comparable with that of the dark matter, spread over the whole halo. It is because of the coincidence that the factor of one-tenth turns up in these two contexts that the rotation speed in the inner regions, where the bright stuff dominates, joins smoothly onto the rotation of the outer parts, where the dark matter of the halo provides the main gravitational force.

The other main type of galaxy—ellipticals—also have dark haloes. The evidence for this is of a different kind. Gas that is held in the gravitational grip of such a galaxy gets hot enough to emit X-rays, and the strength and spectrum of the X-radiation from different regions in and around an elliptical galaxy (the "X-ray profiles") reveal the nature of the gravitational field out to large distances. This indeed suggests that the haloes of giant ellipticals may be even heavier than the haloes around disc galaxies. Because ellipticals do not rotate much at all, the rotation test is of no use here. But some ellipticals are surrounded by faint shells of material, thought to be made of stars from a companion galaxy that has been totally disrupted and is now being swallowed up by the elliptical. The pattern of these shells around the central galaxy depends on the amount of matter causing the disruption and the way it is distributed; this corroborates the X-ray evidence that such a system may contain more than ten times the mass visible in bright stars.

All of these studies are borne out by investigations of the ways galaxies move in binary systems and larger groups. In clusters, the dark matter is torn off the individual galaxies by tidal forces and fills the whole cluster. There is no evidence that conflicts with the idea that 90 percent of the gravitating mass is dark, and plenty that supports it. There is no longer room to doubt that dark matter holds our Galaxy, and others, together; the major issue now is, what is that dark matter?

Dusky Dwarfs

Bright matter makes up only about 1 percent of the density needed to make the Universe flat. Big Bang calculations say that as much as 10 percent of the mass needed to make the Universe flat, perhaps even 20 percent, is allowed to be in the form of baryons. So there is plenty of scope for dark baryonic matter to be lying around in our Galaxy, even though exotic particles are needed to complete the job of flattening the Universe.

There may even be some dark stuff in the thin bright disc of a galaxy like our own, the "white" of the "fried egg." By studying the way stars move on a local scale, in our part of the Milky Way, astronomers can see the gravitational influence of extra mass, holding stars tightly confined in the plane of the disc and not allowing them to bob very far up and down out of the plane. Estimates of how much dark matter there may be in the disc vary but there may be up to twice as much as that in the stars we see.

Could our galactic disc, perhaps, contain large numbers of dim stars, too faint to be seen even with the aid of telescopes from Earth? Such stars are known as "brown dwarfs," and a report of the discovery of what may be the first identified brown dwarf appeared late in 1987. Benjamin Zuckerman, of the University of California at Los Angeles, and Eric Becklin, of the University of Hawaii, used infrared techniques to search for the faint glow of radiation from any brown dwarf that might be orbiting known white-dwarf stars. Most stars occur in binary systems, so it is logical to look for another star alongside a known star. But there is no point in looking for a brown dwarf alongside an ordinary bright star like our Sun, because the sensitive detectors needed to see the brown dwarf would be dazzled by the light from the bright star. So, the obvious place to look is indeed next to a known, faint star.

A white dwarf known as Giclas 29-38, about 45 light-years away from us, turns out to be a source of just the kind of infrared radiation expected from a nearby brown-dwarf companion. A star slightly bigger than the planet Jupiter (and about sixty times more massive) would fit the bill neatly, with a surface temperature of about 1,200 degrees C.

There may be other explanations of the infrared radiation from Giclas 29-38. The most daring speculation would be that what we are seeing is the waste heat from a civilization that has grown to cover the entire surface of a planet. This is not very likely, unfortunately, since on its way to becoming a white dwarf a star swells up into a red giant, engulfing any habitable planets it may possess, before shrinking inwards upon itself—not the sort of conditions conducive to the evolution of an advanced civilization. A cloud of inorganic dust enveloping the white dwarf could also account for the infrared radiation. For the moment, the brown-dwarf interpretation looks the best bet, but it would be nice to find some more brown dwarfs to back up the supposition.

That backup may already have emerged. David Latham, of the Smithsonian Astrophysical Observatory, presented evidence of just such a discovery to the August 1988 meeting of the International Astronomical Union, held in Baltimore, Maryland. Although we have been cautious about including too many mentions of hot news in this book—hot discoveries have a way of cooling off very quickly—this one is so solid that it merits a mention. During a survey of faint stars, Latham and his colleagues found one, HD 114762, whose spectrum shows a periodic Doppler shift over a range corresponding to a velocity of 0.5 kilometre per second, over a period of 84 days. The most likely explanation of this is that the star is being tugged by a large companion, at least ten times more massive than Jupiter, in orbit around it—an orbit more like that of Mercury (which orbits the Sun once every 88 days) than like that of Jupiter. The exact mass of the companion depends on how we are viewing the orbit, which is not known—if it is edge-on, the mass is about ten times that of Jupiter, but if we are seeing the system face-on then the companion is much bigger and certainly ranks as a small star. Whether you regard something with ten times Jupiter's mass as a star or a planet is a moot point. We would call it a brown dwarf, but some people find it more exciting to describe the find as the first definite discovery of another planetary system.

Brown dwarfs are interesting in their own right, as objects intermediate between a star like our Sun and a large planet, like Jupiter. Jupiter has a diameter 11 times bigger than that of the Earth, and its mass is 318 times that of our planet. But that is still only 0.1 percent of the mass of the Sun. A ball of gas even 10 times more massive than Jupiter, with 1 percent of the mass of our Sun, never gets hot enough inside for nuclear fusion to begin to turn it into a star; one with 8 percent of the Sun's mass "burns" nicely. In the intermediate range, brown dwarfs with masses between 1 percent and 8 percent of the Sun's mass fizzle quietly. If there are many brown dwarfs, including some that wander alone through space, they may well contribute significantly to the mass of the disc of our Galaxy. Some of this disc stuff might also be in the form of smaller objects, lacking enough mass to become even brown dwarfs—"Jupiters," balls of cold gas like the giant planets of our Solar System. And some may be in more ephemeral form still, like the comets that occasionally pass through the inner regions of the Solar System in a blaze of glory, and that must come from deep space, although nobody knows exactly where they originate. But it is when we look beyond the disc of a galaxy that we begin to come to grips with the true stuff of the Universe, for the mass of the disc is much less than half the mass of the galaxy.

Black Hole Beasts

The nature of the halo stuff may be connected with a long-standing astronomical puzzle—where are the first stars? All the stars that have ever been studied contain at least traces of elements heavier than hydrogen and helium. In a cavalier generalisation (justified by the fact that 99 percent of all baryonic matter is in the form of hydrogen or helium), astronomers call everything else "metals." Metals can be made only in stars, and are then distributed through the Galaxy by supernova explosions. The first stars must have contained no metals at all, and even if they produced them in their deep interiors, their surface layers should still show no trace of metals. Only later generations of stars, produced from clouds of gas and dust laced with the debris of supernova explosions, should show traces of metals in their spectra. But nobody has ever found a zero-metallicity star.

For historical reasons, astronomers call stars like our Sun, members of the disc of our Galaxy, "Population I." These are the youngest stars, containing most metals. *Older* stars, typical of the globular clusters that move through the visible halo of the Galaxy (but which is far

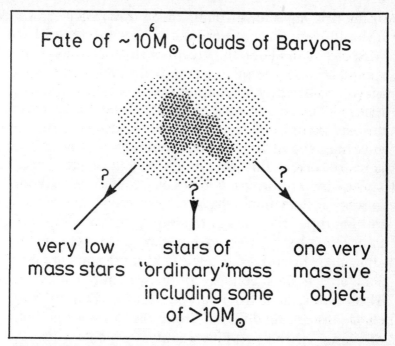

Figure 5.4 In most cosmogenic schemes, the first stars form in collapsing clouds of around 10^6 solar masses. Should we expect a single supermassive star, a small number of very massive objects (VMOs), or a population of ordinary (or perhaps even very low mass) stars? This question relates crucially to the nature of the dark matter, the origin of the first heavy elements, and the background radiation. But we are too ignorant of the relevant processes to give a reliable answer.

less significant, we now know, than the dark halo) are called "Population II." They show only traces of metals. So, naturally, the "missing" zero-metallicity stars, made from the original baryonic stuff of the galaxy, are called "Population III"—even though they have never been detected.*

Many Population III stars must, of course, have run through their life cycles and exploded, or we would not be here worrying about them. Our bodies are largely made out of "metals" manufactured inside stars. Some

---

*These are the stars that may be responsible for the bump in the cosmic background spectrum discussed in chapter 3.

of the first stars have simply faded away more quietly into oblivion. But any ordinary stars that were born with less than about 80 percent of the mass of our Sun should still be around, burning their nuclear fuel sparingly, and brightly proclaiming their presence and their metal deficiency. It must be that the first stars to form were *all* big, unstable systems that exploded very quickly, enriching the interstellar environment, and were then gone from view. But what might they have left behind?

The obvious answer is black holes. Individual Population III stars might have masses anywhere up to a million times the mass of the Sun. Stars heavier than a hundred solar masses do not explode as supernovae. Once their brief lives as nuclear burning stars are over, such objects must collapse all the way into black holes. These holes, made from baryonic material after the Big Bang, are quite different from the primordial holes discussed in the previous chapter, which could have any mass. But, in spite of the popular image of a black hole as a seething maelstrom of violent activity, these could be floating around the halo of our Galaxy in large numbers, quite undetected.

A black hole becomes conspicuous only when there is matter falling into it—when, if you like, it has something to eat. An isolated black hole is very difficult to detect, because it emits no energy (unless it is a minihole in the process of exploding, which these objects certainly are not). It can only be detected by its gravitational influence (which, as we shall see in chapter 8, is difficult but may not be impossible). But such beasts are messy eaters. If a black hole does find something to eat, it does so in a very visible fashion. Such a hole passing through a cloud of gas and dust would suck matter in because of its gravitational attraction, and the infalling gas would pile up around the black hole as it tried to funnel across the Schwarzschild surface. More matter falling onto this piled-up gas would make it hot, and the whole system would radiate energy.

Unfortunately, nobody has been able to calculate exactly what kind of energy an accreting black hole like this would radiate, so nobody knows what kind of spectroscopic "signature" we should be looking for. Since very massive holes would be likely to produce correspondingly large outbursts, however, the fact that nobody has yet seen anything that might be interpreted as such an outburst tells us that it is unlikely that any such objects in the halo of our Galaxy have an individual mass of more than a million Suns.

Stars heavier than the Sun, which formed when the Galaxy was young, would have died by now. Only the remnants of *very* massive stars (well above 100 solar masses) are serious dark matter candidates. Stars of more moderate mass, which die as supernovae rather than by collapsing, cannot contribute much mass in the range up to a hundred times the mass of the sun, because if there had been many of them around they would have left far more heavy elements, including carbon, nitrogen, and oxygen, than are seen in stars or gas clouds in space. At the other extreme, the halo stuff could, however, consist of brown dwarfs or Jupiters.

There is no reason why low-mass stars should not have formed in the halo, and they would be very hard to spot, because they are so faint. But is there any reason to think objects like this *did* form when the Galaxy was young? The best evidence comes from studies not of our Galaxy or any other individual galaxy, but of the hot gas that has been identified flowing into the inner regions of some *clusters* of galaxies.

Baryons May Be Cool

If the dark stuff of the halo is mainly in the form of low-mass stars, then something special must have happened to convert baryonic gas into such objects without producing enough stars like our Sun to make the halo visible from Earth. As far as we know, no such process is at work in the Galaxy today. When we look at regions where stars seem to be forming, such as the cloud of gas and dust known as the Orion Nebula, we see no evidence that low-mass stars are preferred. Brown dwarf halo stars, if they exist in large numbers, must, compared with stars that form today, represent a "special creation" of objects formed in a different way. But this is no reason to reject the idea, since after all the Galaxy was a very different place when those stars were forming than it is today—which is where comparisons with gas in galaxy clusters comes in.

Many clusters of galaxies produce electromagnetic radiation at X-ray frequencies. These X-rays do not come from the stars and galaxies of the cluster but from hot gas between the galaxies. This gas is flowing inwards towards the core of the cluster, where there is usually a massive, dominant galaxy. The gas gets heated by compression as it flows inwards, because gravitational energy is converted into thermal energy, but this energy is dissipated in the form of radiation. Although the gas *is* hot, and that is why we can see it, it is *losing* energy, radiating it away into space. So, such an inwards flow of gas is known as a cooling flow.

In a cooling flow of this kind, there is so much gas that, if the flow had been running ever since galaxies began to form, it could have provided all of the mass of the central galaxy. But if the material we see cooling was being converted into stars in the same way as stars form today in the neighbourhood of our Sun, producing the same mixture of large and small stars, the central galaxy would be very blue and bright. In fact, this is never the case. The cooling flow must be going somewhere—it cannot get back out of the cluster, because it is trapped by gravity. It is not going into the form of hot, blue stars. So it seems that the only possibility is that the material is forming millions upon millions of small, cool stars—brown dwarfs, or, perhaps, Jupiters. There are also cooling flows associated with some individual galaxies, not in the hearts of clusters—probably, dark matter is similarly accumulating in those galaxies.

When our Galaxy was young, it consisted of a very large cloud (or clouds) of gas, falling in towards a common centre under the pull of gravity. Gas flowing into the centre in this way would have been in an environment much more like that of a cooling flow than like the disc of the Galaxy today. It would be surprising if the stars that formed under these different conditions were the same as each other, but it would not be at all surprising if the stars that formed in the young Galaxy were like the ones that are now forming in cooling flows—faint, dim stars.

This suggestion runs against a trend among astrophysicists today to "explain" *all* of the dark matter in the Universe in terms of exotic particles. But visible matter, after all, comes in many different forms, including large and small stars, planets, people, and clouds of gas in space. Why shouldn't dark matter come in many different forms? Either brown dwarfs or black holes could form the dark halo of a galaxy like our own and provide the "missing mass" needed to hold clusters of galaxies together gravitationally. As things stand there is no reason why any single option must explain all of the dynamics of galaxies. On the other hand, axions or other cold dark matter particles *could* do the job very nicely on their own.

When data are sparse, and are compatible with several different hypotheses, we prefer to remain agnostic rather than dogmatic. (Some cosmologists are justly chided for being "often in error but never in doubt.") Future observations and experiments should before long narrow down the range of options. In the meantime, the right strategy for theoretical astrophysicists is to explore all possibilities that seem credible. Further work may suggest new types of observational tests or even reveal unforeseen inconsistencies in some models, thereby narrowing the field. Some individual theorists are happy to explore two (or more) mutually contradictory ideas simultaneously. Others fall in with a particular hypothesis, which they defend and explore to the exclusion of all others (though they may transfer their affections later to a rival, if their initial favourite is discredited beyond all hope of restoration). Either way, the *collective* efforts of theorists should complement those of observers in gradually bringing our knowledge into sharper focus.

Making Mountains out of Molehills

We have already, in chapter 4, run through the list of possible nonbaryonic candidates to provide the dark matter in the Universe, and any of those candidates could be the dominant component of the halo, or could share the halo with one or the other form of baryonic dark matter. Physicists may soon firm up their ideas on what particles should exist. Better still, some form of "ino" might show up in the laboratory. Then, astronomers would have a much better idea of where to search for the missing mass. As yet, however, we have only the astronomical observations with which to judge the plausibility of different candidates as explanations for the presence of massive haloes around galaxies. What kind of tests might show that these galactic mountains are, in fact, made up of ino molehills?

Heavy particles, each with a mass comparable to that of the proton (such as the hypothetical lightest supersymmetric partner), could have an influence on the way individual stars evolve. If the massive halo is composed largely of this kind of ino, sometimes referred to as Weakly Interacting Massive Particles, or WIMPs, then a star like the Sun should gather them up as it orbits around the Galaxy. Over the lifetime of the Sun, perhaps as much as one-trillionth of its mass (10^{-12} of the solar mass) could have built up in the form of WIMPs trapped in its core by gravity and bouncing around among the protons and neutrons. The effect of such a WIMPy core on a star would be to lower the temperature at its centre, because in jostling against the protons and neutrons the WIMPs would spread the warmth at the heart of the Sun out over a broader region. The maximum temperature inside would be lower, but the region over which high temperatures existed would be bigger, so the total output of energy would be the same.

Solar astronomers are interested in this possibility because it could help them to explain a long-standing puzzle—that our Sun produces fewer neutrinos than standard models, with a very high temperature at the centre, predict. If the central temperature were a little lower, as WIMP models require, then fewer neutrinos would be produced, in line with the observations. There would also be implications for other stars.

The WIMPs that would affect stellar evolution come in the category of cold dark matter (described in chapter 3). Computer simulations of galaxy formation provide independent indications that galactic haloes are indeed made from CDM. These simulations can tell us how galaxies would be distributed—whether, for instance, they lie in sheets and filaments—in the different models. They can also show how individual galaxies grow against a CDM background. The background might be axions, photinos, gravitinos, or something else—the models are not fussy at that level. When the computer modellers set their parameters to provide a background sea of CDM in such a way that galaxies form only in the highest peaks of density fluctuations (the "biasing" mentioned in chapter 3), structure very like the geography of the real Universe emerges from the simulations. Once these conditions have been selected to produce a broad picture like that of the real Universe, the modellers can fine-tune the computer programs to focus on the equivalent of a small region of space, containing about ten simulated "galaxies," and watch their evolution in detail. All the features of real galaxies seem to arise naturally in these simulations. Although it is CDM that dominates gravitationally, what we see is the baryonic component. To predict what a galaxy looks like, therefore, requires still more complex computations of how gas clouds interact and form stars. The gas-dynamics calculations suggest that central bulges form early, as gas flows into the central region of a high-density fluctuation and forms stars; discs grow around the central bulges as orbiting gas settles down around the central condensation; and the whole system is automatically embedded in a massive halo of cold dark matter. Faint ellipticals arise when two haloes in the process of formation combine, and big, bright ellipticals result when galaxies that have already formed collide and merge.

This is not definitive proof that halo stuff is cold dark matter, but it is persuasive circumstantial evidence. The important point is that by setting up their computer models to produce a broad picture resembling the sheets, voids, and filaments of the real Universe, and with just 10 percent of the mass in the form of baryons, galaxies with massive haloes and the rest *automatically* emerge from the calculations. Is this merely a coincidence? Or is it a discovery with cosmic significance? The dark matter is unspecified; it could be very light particles, like axions, or it could be more massive WIMPs. WIMPs interact so weakly with ordinary material that they pass easily through the atmosphere, and through the walls of any building. If they constitute the dark matter in our Galaxy's halo, there are 10,000 of them in every cubic metre around us, swarming through any laboratory at speeds of up to 300 kilometres a second. Most of them would go straight through any laboratory-scale object. But somewhere between 1 and 1,000 WIMPs per day (the number depends on details of their properties, which are not yet known) would interact with a kilogram of detector, which contains about 10^{27} baryons. An individual interaction of this kind, where a single atom of everyday matter recoils from the impact of a WIMP, could be detected, if it occurred in a very cold solid. If the dark matter were axions, however, there would be less chance of detecting them in the laboratory, although it would still be feasible.

Ingenious schemes for detecting a halo population of exotic particles seem among the most worthwhile and exciting high-risk experiments in physics or astronomy—potentially as important as those that led to the discovery of the cosmic microwave background in the 1960s. The risk is simply that of failure—there is no danger involved for the experiments. A null result—failure to detect exotic particles—would surprise nobody; on the other hand, such experimenters could reveal new supersymmetric particles (or axions, as the case may be), as well as determining what 90 percent (or more) of the Universe consists of. These experiments could best be done underground, to reduce the background noise caused by cosmic rays—another illustration of how dramatically the scope of observational astronomy has expanded since the days when our knowledge derived solely from optical telescopes. Because the "signal" recorded in such detectors would be sensitive to the velocity of the incoming particles, it would vary over the course of a year, as the Earth moves around the Sun. Such an annual modulation, with a variation of a few percent and a peak in June, when the Earth is moving fastest relative to the halo, would be an unambiguous signature of halo CDM.

More Answers than Questions

The difficulty is not finding a candidate for the dark stuff in the haloes of galaxies but choosing among the many good candidates we have. The halo stuff could be baryons, either in the form of large black holes formed after the Big Bang or of Jupiters and brown dwarfs formed in cooling flows as the Galaxy itself formed. Or it could be nonbaryonic cold dark matter, anything from axions to primordial black holes. As far as our Galaxy alone goes, the odds on each of the three main options must be roughly equal. However, if the Universe is indeed flat, the standard Big Bang requires that there must be nonbaryonic stuff about; moreover, massive haloes arise naturally in computer simulations of the CDM cosmology. So, taking a broader perspective, we would expect at least part of the mass of the halo to be nonbaryonic cold dark matter. And within that framework, the way gas would stream into the forming galaxy from outside is so reminiscent of cooling flows seen today that it would be astonishing if low-mass stars were not formed in the process.

If we were forced to come down off the fence, then, we would suggest that the halo of our Galaxy is probably dominated by nonbaryonic cold dark matter, with the Lightest Supersymmetric Partner not only the best candidate but the one most likely to be detected on Earth (and inside the Sun!). Within that CDM background, however, there should also be a significant amount of baryonic material, in the form of low-mass stars or large, planetlike objects. Beyond the dark halo, if the Universe really is flat, the cold dark particles must dominate. In a sense, the halo stuff marks a transition zone between the region where CDM dominates and the region of bright stars where baryons come into their own. And the ultimate expression of baryonic matter lies at the hearts of galaxies, where huge masses are concentrated into small volumes, producing black holes in an altogether different league from anything we have considered so far.

CHAPTER SIX

★

Core Stuff

BIG BLACK HOLES beat at the hearts of galaxies. A suspicion that first emerged in the 1960s, with the discovery of quasars, was confirmed in the second half of the 1980s, when observers developed techniques accurate enough to measure the speeds of stars in their orbits near the centres of galaxies beyond the Milky Way. One implication is that our own Galaxy also has a black heart, a high-density region from which stars that pass near the black hole might be ejected by a gravitational slingshot effect, hurtling outwards from the nucleus of the Milky Way at thousands of kilometres a second—speeds far in excess of the escape velocity from the Galaxy.*

The concept of a black hole is so familiar today, even to the general public, that it hardly comes as a surprise to discover that astronomers were thinking along these lines, invoking supermassive concentrations of matter in galactic centres, more than a quarter of a century ago. It may, however, come as a surprise to learn that these speculations—regarded as very wild and uncon-

---

*The escape velocity is the speed an object has to move to escape from the gravitational clutches of its "parent" body. An object moving slower than escape velocity always falls back whence it came.

ventional at the time—were being made several years before the name *black hole* was first applied in an astronomical context (by John Wheeler, of Princeton University) in 1968. But although the name for the phenomenon may be less than a quarter of a century old, the concept goes back more than two centuries, to the work of the British pioneer John Michell, an underappreciated polymath of eighteenth-century science. It was Michell who, in 1767, pointed out that there are far too many close pairs of stars seen in the night sky for all of them to be a result of a chance alignment of a nearby star and a more distant star along the line of sight. Some of these stars, he reasoned, must be genuine binary systems, two stars orbiting around each other, locked in a gravitational embrace.* This was a key insight in astronomy. By studying the orbital motions of such binaries, astronomers are able to deduce the masses of stars in binary systems, and thereby to infer the masses of similar stars (with the same colour and brightness) that are not part of a binary system. Michell devised a method of calculating the distances to the stars and also studied earthquakes, among other things. But his most remarkable achievement, looking back with the benefit of two hundred years of hindsight, was surely his work on dark stars.

A Brief History of Black Holes

Michell realised that because the speed of light is finite, and because the escape velocity from the surface of a massive object increases if the size of the object is

---

*Arp's work, mentioned in chapter 2, is based on a similar argument; in that case, however, the statistics of the alignments of galaxies and quasars are still a subject of hot debate.

increased but its density stays the same, there must come a point where the escape velocity is so large that not even light can emerge from the "star." In 1784, he wrote in a paper published in the *Philosophical Transactions of the Royal Society:*

> If the semidiameter of a sphaere of the same density with the Sun were to exceed that of the sun in the proportion 500 to 1, a body falling from an infinite height towards it, would have acquired at its surface a greater velocity than that of light, and consequently supposing light to be attracted by the same force in proportion to its "vis inertia" with other bodies, all light emitted from such a body would be made to return towards it, by its own proper gravity.

Michell was not alone in thinking along these lines in the late eighteenth century, although he was the first to express these ideas in print. The Frenchman, Pierre Laplace, independently came up with the same notion some ten years later, and included a discussion of "corps obscurs" in early editions of his masterwork *Exposition du Système du Monde.* By the fifth edition, however, all such references had disappeared, perhaps because Laplace had second thoughts, or perhaps because his colleagues had scoffed at the idea. More than a century was to elapse before the concept of "des corps obscurs" was to surface once again—and when it did, it suffered the same fate, at first, as the ideas of Michell and Laplace.

Two revolutions, in the first decades of the twentieth century, paved the way for black holes. The first was Einstein's general theory of relativity, explaining gravity through the bending action of matter on spacetime. This led, among other things, to Schwarzschild's work, mentioned in chapter 4, which physicists and astronomers largely ignored. The second was quantum physics, which made it possible to calculate the ultimate fate of a star when its supply of nuclear fuel was exhausted. This was seized upon by some astronomers, but with unexpected results. It was by thinking about the "end point" of stellar evolution that a young Indian student, Subrahmanyan Chandrasekhar, "discovered" black holes in a very different context from the discovery made by Michell and Laplace.

The eighteenth-century pioneers had imagined adding more mass to the Sun but keeping its density constant—in effect, packing a hundred million Suns alongside one another in a huge sphere, like a bag full of marbles. Another way to imagine causing an increase in the escape velocity from the surface of a star, however, is to pack the *same* mass within a smaller volume—to make (or let) the star shrink. As the star contracts, its density increases, escape velocity gets larger, and it is harder for light to escape from its surface. General relativity describes this behaviour quite well, and predicts that light from a very dense star will be redshifted, because it has lost energy (even though its velocity cannot change) in the struggle to climb out of the star's "gravitational well." Just such a gravitational redshift can indeed be seen in the light from some dwarf stars.*

Chandrasekhar thought about the problem during the long voyage from India to England, where, in the early 1930s, he was to begin working with the great astronomer Arthur Eddington. He realised that quantum forces involving electrons could hold a star up against the inward pull of gravity only if the mass of

---

*Einstein also told us that the speed of light is constant, so the problem is a little different from the one considered by the eighteenth-century savants. If you imagine the gravitational field of a star getting stronger and stronger, light still escapes from the star, but is increasingly redshifted—increasingly feeble. The point at which escape velocity is equal to the speed of light is the point at which the redshift removes *all* of the energy in the light, so that nothing escapes.

the star was less than about 1.4 times the mass of our Sun. No white-dwarf stars bigger than this limit have ever been found, confirming his calculations. But what happens to stars with more mass at the end of their lives, when nuclear fuel is exhausted? "A star of large mass," Chandrasekhar wrote, "cannot pass into the white-dwarf stage, and one is left speculating on other possibilities."

His mentor, Arthur Eddington, was not impressed. He commented that such a star, according to Chandrasekhar's calculations, "has to go on radiating and contracting and contracting until, I suppose, it gets down to a few kilometres' radius when gravity becomes strong enough to hold the radiation and the star can at last find peace." You might think, from those words, that Eddington deserves credit for inventing the modern concept of stellar mass black holes; alas, however, he was merely pointing out what he saw as the absurdity of Chandrasekhar's notion, and went on, "I think that there should be a law of Nature to prevent the star from behaving in this absurd way."

What seemed absurd to Eddington, however, may be the most natural thing for the Universe. Chandrasekhar, as befitted a young student in a strange land working with the foremost astrophysicist of his day, was suitably crushed, and went on to other work.* But others were less inhibited. In 1931, just a year after Chandrasekhar first presented his analysis of the ultimate fates of stars, the great Soviet physicist Lev Landau showed that stars with about 1.5 times the mass of our Sun must collapse beyond the white-dwarf stage, with electrons and protons being crushed together to make neutrons and the star becoming a neutron star. Even

*He has, however, rounded off a glittering career, sustained for more than fifty years, by returning to the study of relativity and producing a massive treatise, *The Mathematical Theory of Black Holes.*

neutrons, however, would be crushed out of existence if the star had a little more mass still, and it would collapse, the equations said, to a point.

Like Eddington, Landau thought this conclusion was ridiculous, and argued that there must be some law of nature, as yet unknown, that would prevent the ultimate collapse. In the United States, however, Robert Oppenheimer (who was to become famous as the "father of the atomic bomb") took the idea at face value. Working first with George Volkoff, and later with Hartland Snyder, in the late 1930s and early 1940s Oppenheimer developed a mathematical description of the collapse of stars into what we now call black holes. Nobody else seems to have taken the idea too seriously; it seemed like a mathematician's trick, a byway of general relativity with no relevance to the real world. Even if black holes existed (which few believed), they could never be seen, so why bother with them? And besides, many researchers, like Oppenheimer, were soon diverted into work on more pressing problems than the ultimate fates of stars.

Little more was heard of stellar-mass black holes until the late 1960s. Only mathematicians toyed, occasionally, with the Schwarzschild variation on Einstein's theme; and the mathematicians seemed unaware of the attempts by Chandrasekhar, Landau, and Oppenheimer to put these mathematical abstractions into the real astronomical world. In 1968, however, the radio sources known as pulsars were discovered, and quickly explained in terms of rotating neutron stars. It was like a shock wave in astronomy. Neutron stars really existed! If so, then Landau's calculations were correct; and if Landau's calculations were correct, then, as Oppenheimer had shown, stars much bigger than the Sun must collapse completely, beyond the neutron-star state, into black holes.

In the 1970s, new observations using X-ray detectors

placed in orbit around the Earth, above the obscuring blanket of the atmosphere, revealed that the Universe at large, including our Galaxy, is a far more energetic place than anyone had dreamed. Many intense sources of energy, pinpoints in the sky that blast X rays out into space, have now been identified. One of the best ways to make energy is to drop matter into an intense gravitational field. It gets hot, and the hotter it gets the higher the frequency of radiation it emits, up to and including X and gamma rays. The best explanation of many of the X-ray sources now known to litter the Milky Way is that they are sites where a black hole of the kind Eddington dismissed as nonsense is in orbit around an ordinary star, stripping gas off from its companion by tidal forces and swallowing it down its Schwarzschild throat. But these are not the black holes at the hearts of galaxies; those, ironically, are much more like the sort of thing John Michell had in mind. They too, however, became respectable because of new observational developments that showed energy being produced in the Universe on a scale that could be explained no other way.

The Quasar Connection

"Active" galaxies were first identified by an American astronomer, Carl Seyfert, in 1943. He studied what seemed to be ordinary disc galaxies and found that some of them had very bright central cores. Studies of the spectral lines in light from the centres of these galaxies showed that the gas within a few tens of light-years of the centre itself was in a disturbed state; several hundred of these objects, dubbed "Seyfert galaxies," are now known.

Thanks to the opening up of the electromagnetic spectrum through radio astronomy in the 1950s, and space technologies in later decades, the astronomers' view of the cosmic scene changed dramatically in the years following the Second World War, and those Seyfert galaxies, which seemed so unusual and rare in the 1940s, are now a familiar feature of the cosmological picture. It was indeed radio astronomy that provided the first firm clues that there is more to galaxies than the bright stars we can see with optical telescopes and the associated gas between the bright stars. In 1954, a strong source of radio noise from space (known as Cygnus A because it lies in the direction of the constellation Cygnus) was identified with a distant galaxy that had a redshift of 0.05—a high figure by the standards of redshifts known in the mid-1950s. So much radio energy could be detected coming from this galaxy that, astronomers realised, other galaxies might be "visible" to radio telescopes even if the galaxies were so far away that the light from their hundred billion or more stars could not be seen by optical telescopes. And as the radio observations improved, the observers discovered that the intense radio noise from Cygnus A does not, in fact, come from the galaxy itself, but from two "lobes" symmetrically placed on either side of the galaxy. This double structure, in which the two radio components may be separated by a million light-years, or even more, is now recognised as a characteristic of the strongest cosmic radio sources. Something in the central galaxy beams energy out into space across the light-years and causes an interaction with intergalactic material that produces the radio noise.

Radio astronomers began to seek out and collect data on strong radio sources, whether or not they could be identified with galaxies. A new twist in the story came on the last day of 1960, when Allan Sandage reported, to a meeting of the American Astronomical Society, the discovery that one of the radio sources logged in a survey carried out at Cambridge University (the 48th source logged in the third such survey, hence the name 3C 48) could be identified with a bright object that looked like a star, not a galaxy. The "star" lay in exactly the place on the sky that the radio noise was coming from, but nobody knew how it was making the noise.

Such radio stars remained a puzzle until 1963. Then, another source in the same catalogue, 3C 273, was identified with a starlike object. Cyril Hazard and his colleagues in Australia pinned down the source's position with high precision. Maarten Schmidt, at Mount Palomar in California, then found that this "star" had a spectrum that included recognisable features due to hydrogen—but these features were shifted by a prodigious amount, nearly 16 percent, towards the red end of the spectrum. The fact that 3C 273 had a redshift of 0.158 was a valuable clue; the nature of the spectrum of 3C 48 was soon also explained as due to an enormous redshift, in that case 0.368. Quasars had arrived on the astronomical scene.

The name quasar derived originally from the acronym QSRS, which stood for the term quasi-stellar radio source, an apt description of objects like 3C 273 and 3C 48, which look like stars in optical telescopes but produce enormous amounts of radio energy. Many similar objects are now known; they look like stars, emit a great deal of energy, and have high redshifts, but they are not all sources of radio noise. Since quasar is also a perfectly good contraction of the term quasi-stellar, we shall stick with the rule that a quasar is anything, radio-noisy or not, that looks like a star but has a redshift that places it far beyond the confines of the Milky Way.

This, of course, is what all the fuss is about. Quasars look like stars, superficially, but have huge redshifts. If those redshifts are interpreted in the standard way as due to the expansion of the Universe, then we see quasars far away across space and time. In order for them to be visible at all, they must be pouring out huge quantities of energy—and studies of the way the light from quasars varies soon showed that the energy must be coming from a region of space no bigger than our Solar System. Other explanations of the redshift were tested, more or less in desperation, and found wanting. Could quasars really be stars, shot out from the centre of our own or some nearby galaxy in a violent explosion? Might the high redshifts really be a gravitational effect? Neither explanation works—for gravity to produce the redshifts required, we would have to invoke supermassive black holes, and that, as we shall see, is sufficient to provide the powerhouse of a quasar even if its redshift is cosmological; and if quasars are stars shot out from the hearts of galaxies, why do we only see them moving away from us (redshifted), never coming our way (blueshifted)? These arguments, and others, left few astronomers in any doubt, by the end of the 1970s, that quasars were cosmological.

By then, hundreds of quasars and quasarlike objects had been discovered, and the catalogues contained a whole zoo of objects with a confusing array of names (QSO, QSS, BL Lac objects, blazars, optically violent variables, and more besides). All the more reason for us to stick with the generic *quasar*. At the same time, Seyfert-like activity had been found in many galaxies, and even galaxies that did not show such extreme violence in their cores showed signs of lesser activity. The confusion of names, it was eventually realised, concealed the fact that there is a continuum of activity ranging from quiet galaxies like our own through mildly active galaxies to Seyferts and beyond, with the brightest Seyfert galaxies essentially indistinguishable from low-redshift quasars. It also became clear, through superbly skilful observations, that traces of a surrounding galaxy can actually be seen around some relatively nearby quasars. The implication is inescapable—all quasars lie at the hearts of galaxies, and very many, perhaps all, galaxies have something interesting at their hearts. Whether the something interesting ever turns into a quasar, and when and for how long, became a subject of theoretical scrutiny.

Black Hole Powerhouses

A black hole needs to swallow mass in order to release the energy that powers a quasar—but it needs to swallow surprisingly little compared with the total mass of its host galaxy. Allowing matter to fall into a strong gravitational field is, in fact, the most efficient way to convert mass into energy, apart from annihilating a particle with its antiparticle partner. The amount of energy locked up in a mass m is mc^2, and the maximum amount of energy that could in principle be released by dropping a mass m from an infinite distance into a black hole is almost one-half of this. If only a few percent of the mass-energy, much less than this absolute limit, is actually being processed in this way and released as light, radio waves, X-rays, and the rest, a big black hole needs to swallow only the equivalent of one or two times the mass of our Sun each year in order to shine as brightly as the objects we now see at redshifts of 4 and beyond. For an "efficiency" of about 10 percent, and a lifetime of a few million years, that suggests that about a hundred million (10^8) solar masses of material are locked up in the central engine. This, as it happens, is the sort of black hole envisaged by Michell two centuries ago—it is about the mass of a hole whose "density," defined in terms of the mass contained within the Schwarzschild radius, is equal to that of the Sun, and comparable to that of solid objects on Earth. Whereas stellar mass black holes cannot form until after the matter has been squeezed to supernuclear density, formation of supermassive black holes does not involve these physical complications. Such a black hole could even arise simply because too many stars got too close together in the core of a galaxy.

The mass involved is still very small, by galactic standards. A hundred million solar masses sounds a lot, but the mass of the bright stars in a typical galaxy adds up to a hundred *billion* solar masses, a thousand times greater than the hole mass. In other words, the hole represents just 0.1 percent of the mass of the bright stuff in the galaxy surrounding a quasar. And there is, as we have seen, probably ten times as much dark stuff in the halo of the galaxy, so the hole mass represents only 0.01 percent of the whole mass of the galaxy. Don't look in the hearts of galaxies if you want to find the dominant stuff of the Universe!

It is easy to imagine ways in which such cosmically modest black holes can form in the central regions of galaxies. Massive gas clouds may collapse directly into a black-hole state, or they may form dense clusters of stars that merge together and are steadily eaten up by the first stellar mass black hole to form. Matter will inevitably settle towards the centre of a forming galaxy, simply because this is the bottom of the gravitational well; the ultimate fate of large amounts of matter settling into one place must be, by one route or another, to form a massive black hole.

Many quasars and radio sources show visible jets of material streaming out from the central object, as well as the double-lobed radio structure. This is simply explained in terms of the processes that power the quasars. In order to produce energy, a black hole must be gathering in matter (and it is probably no coincidence that quasars and active galaxies have more than their share of nearby companion galaxies; these galaxies, or the tidal distortions they cause in the active galaxy itself, are presumably responsible for matter spilling

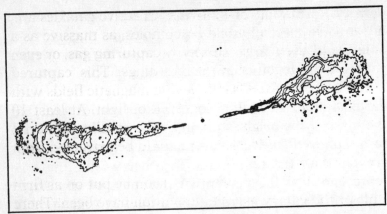

Figure 6.1 This map shows the brightness contours of the radio galaxy Hercules A, whose morphology is typical of strong sources. Energy is being channelled from an active galactic nucleus (black hole?) along narrow jets, and deposited in lobes a million light-years or more from the galaxy.

into the black hole). Any object formed from a collapsing cloud of material must rotate faster as it contracts, just as an ice skater picks up rotational speed by tucking her arms in. A black hole is no exception. So the image of the central powerhouse of a quasar is of a black hole the size of our Solar System, containing the mass of a hundred million Suns, rotating fairly rapidly in the middle of a surrounding cloud of gas, dust, and stars on the point of being torn apart by tidal forces. In such a system, the surrounding material will settle down towards the plane of rotation, as a fat torus, like an enormous doughnut around the black hole. Energy released from the hole as it swallows material would follow the path of least resistance out of this mess, which would be out of the "poles" of the black hole, along the rotation axis of the torus, squirting magnetic fields and plasma (a mixture of atoms stripped of their electrons and the electrons themselves) at a sizeable fraction of the speed of light, out towards intergalactic space.

So the central "prime movers" in active galaxies are, theorists believe, spinning black holes, as massive as a hundred million Suns, fuelled by capturing gas, or even entire stars, from their surroundings. This captured debris swirls downwards, carrying magnetic fields with it and moving at nearly the speed of light. At least 10 percent of the rest-mass energy of the infalling material can be radiated, and still more could be extracted from the hole's spin.

We hope that these ideas may soon be put on as firm a basis as theories of stellar evolution have been. There is still a long way to go, but if we *can* do this we will have an opportunity to learn from a safe distance, by studying galactic nuclei, whether black holes really behave as Einstein's theory predicts. That theory says that, near black holes, space and time behave in ways that run counter to common sense. Time would stand still for someone who managed to hover or orbit in a spacecraft just outside the hole; such an observer could see the entire future of the Universe pass by while taking a few breaths. But anyone really wanting to explore a black hole should definitely pick one of the monsters at the heart of a galaxy. The Schwarzschild surface there encompasses a region as big across as our Solar System, with densities no greater, near that surface, than the water that comes out of a tap here on Earth. Having passed through the Schwarzschild surface in your spacecraft, you would have several hours for leisured (if not quite relaxed) observations before being stretched by tidal forces and then crushed in the central singularity. Unfortunately, though, there would be no way for you to communicate your observations out of the black hole to interested astrophysicists on the other side of the Schwarzschild surface.

All this assumes that you could guide your spacecraft in carefully at the "equator" of the black hole, well away from the poles, where material is being squirted into space from the region just above the Schwarzschild surface. The structure produced by this beaming of energy and plasma out from the central powerhouse of a quasar stretches, in some cases, across 10 million light-years. Since the material in the jet cannot move faster than light, that means that the central quasar must have been active for 10 million years. On the other hand, it is very difficult to see where the energy could come from to keep a quasar "running" for much longer than this, and at the same time the distribution of quasars and radio sources across the Universe suggests that most of the active galactic nuclei that existed in the past have now died out. The age of a galaxy today is comparable with the age of the Universe, around 10 billion years. The lifetime of a quasar, on the other hand, may be no more than 100 million years, just 1 percent of the age of the galaxy it inhabits. We conclude that dead quasars today outnumber the living ones, and that many earlier generations of quasars may now be defunct. A dead quasar would presumably be a massive black hole now almost quiescent because it is starved of fuel—starved either because it is in a galaxy that has been swept clean of gas or because it has gobbled up all the stars near it. So, on this picture, there ought to be evidence of the presence of supermassive black holes in the nuclei of quiet galaxies, like our neighbours in the Local Group and the Milky Way itself. That evidence is now coming in.

Weighing the Evidence

Over the past ten years or so, several astronomers have tried to find evidence that massive black holes do indeed lurk in the hearts of nearby galaxies. Some have claimed success, but there have always been questions about the interpretation of the data. By the end of the 1980s, however, the weight of evidence began to tilt the balance in favour of the idea that most, perhaps all, galaxies have big black holes at their centres.

Some of the most impressive statistical evidence comes from Wallace Sargent and Alex Filippenko, of Cal Tech. At the time we are writing, in 1988, they are in the midst of a survey of five hundred galaxies, obtaining spectra of the light from the central nucleus of each of these objects. More than 10 percent of the galaxies in this sample show the characteristic features (technically, broad emission lines at the wavelength of hydrogen alpha) that are usually taken as indicating the presence of a massive black hole. By this criterion, each of those galaxies would usually be identified as a Seyfert, the class of galaxies previously regarded as intermediate between "ordinary" galaxies and quasars. In the past, Seyfert galaxies have been identified and labelled as such because the broad emission lines are obvious features of their bright spectra; such emission lines are much more common among faint galaxies than was previously suspected. Because it is very difficult to obtain the necessary detailed spectra of faint objects, these positive identifications still probably represent only the tip of the iceberg; more detailed studies (for example, with NASA's Hubble Space Telescope) are now expected to show this low-level "quasar" activity in virtually every galaxy.

Coming closer to home, some of the latest, and best, evidence comes from studies of two near neighbours to the Milky Way. New observational tests reveal the presence of a hole with mass around 50 million Suns in the biggest galaxy in the Local Group, the Andromeda Nebula (M 31), and one of 8 million Suns in a smaller galaxy known as M 32. Two American researchers, Alan Dressler and Douglas Richstone, carried out a spectroscopic study of the motions of stars in the inner regions of both these galaxies. Because they are relatively nearby (only 2 million light-years away), the study showed details of orbital velocities on a fine scale, in close to the hearts of the two galaxies. These observations, and corroborative ones by John Kormendy (of M 31) and John Tonry (of M 32), are very hard to explain except by the presence of supermassive black holes. And since these two galaxies are otherwise completely ordinary and undistinguished, showing no signs of unusual activity in their nuclei, the implication may be that *all* galaxies harbour supermassive black holes. In which case, we surely ought to be able to find evidence of one in our own Galaxy.

At the Heart of the Milky Way

Conditions at the nucleus of our Galaxy are certainly very different from those in the neighbourhood of the Solar System. Unfortunately, because there is a lot of gas and dust in the disc of the Galaxy, and our Sun is also orbiting in the disc's plane, the central region itself is obscured from view to optical telescopes. Infrared radiation and radio waves can, however, penetrate this obscuring dust to some extent, so it is from radio and infrared observations that we glean an image of the heart of the Milky Way.

There is a very small, variable source of radio noise right at the galactic centre. This source is too small to be resolved by radio interferometers; at the distance of the galactic centre this implies a size measured not in light-*years* but actually less than one light-*hour* (that is, the distance light could travel in 60 minutes) across. For comparison, it takes light from the Sun 160 minutes to reach Uranus, the farthest giant planet in our Solar System. As well as other interesting activity from

this region, there is strong gamma radiation characteristic of the kind produced when electrons and positrons
pairs annihilate one another. That in turn means that
electron-positron pairs are being created by some form
of energetic activity, and accretion of material onto a
black hole with a mass of up to a million Suns would
fit the bill.

There certainly cannot be a *monster* black hole in our
galactic centre. The way gas is moving within a few
light-years of the centre shows that a million solar
masses is just about the upper limit, so the evidence
actually fits well together. *Our* Galaxy has never been
a fully fledged quasar; but the balance of opinion has
now shifted so far in favour of the idea that all galaxies
harbour large black holes that it has become natural to
ask what sort of detectable effects on a lesser scale than
quasar activity such a hole might produce in our own
Galaxy, and others. A little thought soon throws up
some intriguing possibilities.

Jack Hills, an expert on stellar dynamics from the
Los Alamos National Laboratory, in New Mexico, suggests that a million-solar-mass black hole at the centre
of our Galaxy may be spitting out stars moving at 4,000
kilometres a second once every 10,000 years. If so, some
two hundred such fast-moving objects ought to lie within
the radius of the Sun's orbit about the galactic centre,
and some of them ought to be detectable, moving at far
greater than the escape velocity from the Galaxy. The
discovery of one or more such objects on the way out
into the depths of intergalactic space would be proof
that a massive black hole lies at the centre of the Milky
Way.

Hills has investigated the way pairs of stars in tight
orbit around each other ("hard binaries") can be disrupted as they pass by such a black hole. Many stars
occur in binary systems, and where stars are packed
closely together, as they are in the core of the Galaxy,

encounters between stars can transfer orbital kinetic energy from a binary system to a passing star, "winding up" the binary and binding its two stars more closely together.

When such a binary passes near a massive black hole, however, something quite different can happen. Depending on the size of the binary system, its orbital speed, and the speed and angle at which it passes the hole, one star in the binary may be captured by the hole while its companion is ejected at very high speed (the process is reminiscent of, but actually quite different from, the way one member of a virtual pair produced near a minihole can escape while the other is captured).

The stellar density in the heart of the Milky Way is poorly known, but estimates suggest that a star should pass as close to the hole as the Earth is to the Sun once every 100 years. If just 1 percent of the stars involved in these close flybys are hard binaries, one escapee with a velocity of 4,000 kilometres per second should be produced every 10,000 years. Because it is 35,000 light-years from the galactic centre to the distance of the Sun's orbit, it will take each runaway 2 million years to get as far out from the core as we are. In that time, two hundred runaways are spat out, which is why there should be two hundred such superfast stars en route outwards but still within the radius of the Sun's orbit at any time.

Such stars ought to be easy to spot. Even at a distance of 35,000 light-years, the maximum motion of such a star across the sky would be 0.1 second of arc per year. Other hypervelocity stars will be closer, brighter, and more obvious. So why have none been reported in standard sky surveys? It may be that observers have indeed noted such objects but have assumed that they are in fact much closer stars, moving correspondingly more slowly. Since such stars will shoot out from the central black hole randomly in all directions, there is only a very slim chance of one passing so close to us that its nature would be obvious even if you were not looking for something of this kind. If they are now identified in their true colours, they will provide the strongest evidence that our Galaxy harbours a supermassive black hole. On the other hand, the absence of evidence for such objects can never be taken as proof that they do not exist—we may simply be looking for them in the wrong place. Happily, though, there are other effects related to the production of these fast stars that would produce an unambiguous signature in a unique place, at the centres of nearby galaxies themselves.

A Flare for Black Holes

Hills has considered the fates only of *binary* systems moving dangerously close to large black holes. Even single stars, however, may be disrupted in interesting ways, not just swallowed whole, if they venture too close to such an engine of destruction. The dynamics of stars in the inner regions of nearby galaxies such as M 31 and M 32 indicate the presence of central black holes, and since these are nearby, well-studied galaxies we also know a great deal about the number of surrounding stars in each case and the way they move, so we can calculate how often they will be captured by the hole. The best way to test the idea that this is indeed happening is to find an effect that must be produced if the holes are there but cannot be produced by anything else. The best bet may be the distinctive features of the way energy is released when a star is partially gobbled up by the black hole.

A hard pair binary stars in which one star is swallowed and the other escapes is in many ways a simplified picture of what happens to a single star in the embrace of a supermassive black hole. As the star gets close to the hole, it experiences large tidal forces and may be disrupted, losing matter or being broken up completely. Part of the debris is expelled, at speeds up to 10,000 kilometres per second, by a gravitational sling-shot effect; the remaining debris would be left gravita-tionally bound to the hole, and in orbit around it, destined to dribble down the Schwarzschild throat. How long might this take?

For black holes with a mass of only a few million Suns, it turns out that each stellar "meal" is digested long before the next close encounter occurs. The result would be a short-lived flare of activity from the nucleus of the galaxy, an outburst lasting for only a few months or years. Meanwhile, the fraction of the original star that had escaped from the hole would not remain in-tact, but would fan out in a stream of debris mingling with the other material of the surrounding galaxy, and producing no discernible sudden burst of energy.

As far as our own Galaxy is concerned, the fact that we do not see a flare of activity from the centre of the Milky Way today poses no problem, since we would expect such a flare only once every 10,000 years or so; we are much more likely, however, to see such flares in other galaxies. Because of the rarity of such events (in round terms, if each galaxy flares every 10,000 years, we need to study 10,000 galaxies in order to see one flare a year), it is still no surprise that no such flare has yet been seen; but with ever-improving telescopes and knowing now what they are looking for, there is every chance that observers might spot such a flare from a galaxy no farther away than the Virgo Cluster within the next few years.

In our own Galaxy, the outgoing debris, even if it is not in the form of the whole, hypervelocity stars envis-aged by Hills, may be more conspicuous. Material spraying out from the central black hole, like water from a garden sprinkler, would provide a barrier that slowed infalling gas. This would have two effects. It might *reduce* the activity of the hole itself, by reducing its supply of "food," and it would surely produce a hot bubble around the centre of the Galaxy, where the incoming material and the outgoing debris meet head-on. Again, now that the theoretical calculations show what form this bubble might take, observers can begin to look for predicted effects. They may not have to look far. X-ray observations show that there is indeed hot gas in the central 1,000 light-years or so of elliptical galaxies, and if this really is helping to stem the flow of material inwards, then even the quietest of nearby galaxies could be harbouring a semistarved black hole with a mass as big as a hundred million Suns.

These are all exciting prospects, far removed from the everyday world in which gravity is simply the force that causes apples to fall downwards or breaks a skier's bones in a fall. It is, perhaps, worth pointing out that all of the ideas described in this chapter represent mainstream thinking among astrophysicists today. Over the past twenty years, theorists working on this subject have sometimes had the illusion of rapid progress. What we've really had is a rather slow advance, with "sawtooth" fluctuations as fashions have come and gone (what seems like two steps forward being followed by one pace backwards). But the idea that quasars are powered by supermassive black holes at cosmological redshifts, and even the idea that a million-solar-mass black hole lies at the centre of the Milky Way, spitting out stars at speeds of thousands of kilometres per second, are now routine. If quasars had been discovered in, say, 1973, *after* the discovery of pulsars and compact X-ray sources in our Galaxy, and after the resulting theoretical developments, then surely a consensus around models involving massive black holes would very quickly have emerged. It took so long to establish this as the "best buy" simply because, in the early to middle 1960s, no detailed work had been done on black holes for a quarter of a century, and nobody had expected to find big black holes at large in the Universe. By the time the black-hole models were being refined, various speculations had had time to take root and grow, and it took correspondingly long for the black-hole model to catch up and surpass them.

But there is no need to feel disappointed by the news that big black holes are routine physics, far from the cutting edge where speculation can still have free rein. If your taste is for wild-eyed conjecture, look no farther than the next chapter.

CHAPTER SEVEN

—————————★—————————

Cosmic String

IT MAY BE A BACKLASH against the days when exotic names were the vogue both in cosmology (quasar, blazar, supermassive black hole . . .) and particle physics (charmed quark, colour theory, grand unification . . .), but two of the most important ideas in contemporary physics each go under the most prosaic of names— *string*. In terms of scale, the two kinds of string could hardly be more different. To a particle physicist, "strings" are the entities that replace the old concept of particles. Instead of thinking of particles in terms of mathematical points of mass-energy, tiny billiard balls, theorists are learning to describe them in terms of tiny lines or loops of one-dimensional string, far smaller than protons or neutrons. To a cosmologist, on the other hand, "string" may stretch, literally, across the Universe, although even that kind of string is far thinner than a single atom.

The two concepts are related, in that both emerge from the arena of very-high-energy physics and the search for a unified theory of all the particles and forces of nature. But they should not be confused with each other, so it helps that physicists usually refer to each type of string with an adjective attached—"cosmic

175

string," for the kind that stretches across the Universe, and "superstring," for the kind particles may be made of. It's cosmic string that matters when you are worrying about how galaxies form in the first place, and why they lie in sheets and filaments* around the edges of empty bubbles in the Universe. But the approach to unified theory that gives us superstrings also has something to say about the nature of dark matter that may influence the visible Universe by its gravity alone.

A Theory of Everything?

The new theory of superstrings grew out of the search by mathematical physicists for a single theory, one set of equations, to describe all the forces and particles. Their concepts are still tentative, but theorists no longer believe that the quest is hopelessly premature—it is no longer just cranks who try to "solve" all of fundamental physics in one go. Such a "theory of everything," or TOE, must go beyond both quantum physics (which, in the form of quantum chromodynamics, QCD, successfully explains much of the particle world) and general relativity, which deals with the Universe at large and with gravity, the force most difficult to squeeze into a unified theory. But since both those theories work superbly within wide limits, a good TOE must include each of them within itself.

So far, so good. What are the problems that need to be solved? One interesting feature—not really a problem— is a kind of chicken-and-egg puzzle about general rela-

---

*Such filaments have also sometimes been referred to as *strings* of galaxies. This third application of such a prosaic term is too much of a good thing, and we shall stick to *filaments*, or *chains*, to describe the appearance on the sky of long lines of galaxies.

tivity. Starting out, as Einstein did, from a description of curved spacetime, the theory requires the existence of gravity waves, ripples in the fabric of spacetime, and the associated graviton, a particle with zero mass and spin 2. If you prefer, however, you can start out from a theory based on a zero mass, spin 2 graviton; it yields the usual form of general relativity, with curved spacetime. Until recently, there was no reason to regard either view as providing a better insight into the nature of the Universe. But that may be changing.

The big problem with all particle theories prior to string theory is that they lead to infinities when gravity is included. Now, some infinities are embarrassing but can be lived with. QCD, for example, is riddled with infinities, which are swept under the carpet by a trick called renormalisation, and ignored. Renormalisation is essentially just a mathematical ploy. But it leaves a set of equations that can be used, in the right circumstances, to describe how particles behave. When gravity is included, the infinities *cannot* be renormalised, and cannot be ignored. They loom up in the equations and make them impossible to work with.

All this happens when particles are regarded as mathematical points, the simplest entities imaginable. So, more or less in frustration, some theorists decided to see what happened if they considered particles not as points but as the next simplest entities imaginable, little one-dimensional lines, or strings. It turns out that not only does the problem with infinities not arise, but that rather than having to add gravity into the theory, they found it there already—gravity, and specifically a mass zero, spin 2 graviton, is already a necessary part of any workable string theory of the particle world.

Why *super*strings? The adjective creeps in from another particle theory, called *supersymmetry*. This is the theory that says that every type of particle that is

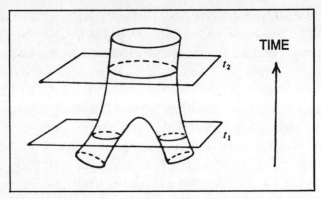

Figure 7.1 The "trouser" spacetime diagram of two super-strings that are separate at t_1 but have merged by time t_2.

associated with a force (like the graviton) must have a partner (the gravitino, in this case) that belongs to the material world, while every particle that we are used to thinking of as, well, a particle (an electron, say) has a partner belonging to the force family (the selectron). Supersymmetry plus string theory gives you super-strings, the best candidate yet for a theory of every-thing, and one that gives one particularly delightful image.

In string theory, particles are represented by little loops (we do mean little—typically about 10^{-33} centime-tres across), which sweep out tunnels through space. When two string loops meet and merge, their behaviour can be represented diagrammatically by a structure with a striking resemblance to a pair of trousers. Space-time trousers, it seems, may provide the ultimate de-scription of the particle world.

But the most important feature of the search for a TOE remains what John Schwarz, of Cal Tech (one of the founders of superstring theory), has described as a "Deep Truth"—that every consistent variation on the string theme has one, and only one, graviton, a mass-less, spin 2 particle that leads inevitably to the spacetime-curvature description of general relativity, and thereby, by implication, on to Newtonian gravity when the gravitational fields are weak. This is a coincidence that simply cannot be ignored. Progress towards the ultimate physical theory can, it seems, be made only by going further along the path trodden by Einstein and by Newton himself.

There is, however, still a long way to go. Although superstring theories agree on the overall nature of strings—a typical length of about 10^{-35} metres and a typical tension in the string that stores energy equivalent to 10^{38} proton masses, numbers related to the Planck mass and Planck length mentioned in figure 3.1—there are many different versions of superstring theory to work with. Some of the most successful work only in ten dimensions, which introduces the problem of explaining why our Universe seems to have only four dimensions, three of space and one of time. The problem is not insurmountable—mathematicians have a ploy called *compactification*, which allows the extra dimensions (six, in this case) to be rolled up so small that they are unobservable. The effect is rather like the way in which a two-dimensional hosepipe, a sheet of material wrapped around a line, looks like a one-dimensional line if viewed from far enough away. Each point of our ordinary space, at each moment of time, is really a tiny but complex curled-up six-dimensional world. The interconnected structures and resonances in these extra dimensions control how the superstrings behave, and thereby determine what particles exist at each point in ordinary space, and how they interact.

There are different versions of ten-dimensional superstring theory, different versions of compactification, and even versions of string theory that start out in four dimensions. Superstring theories are not mathematically elegant (they don't hang together naturally in the way that, say, general relativity does), which is taken as a sign that the "right" mathematical description has not yet been found; nor are they based on some deep truth, such as the geometrical principles that Einstein used as the foundation of general relativity. Schwarz himself says that it is unrealistic to expect too much too soon, and that although the way superstring theory does away with infinities and brings gravity into the fold of unified theory is striking and encouraging, it is likely to take a few decades of hard work before there is a really satisfactory understanding of what superstring theory is all about. (Unlike most physical theories, which have made use of a "mathematical language" that was already developed, superstring theory presents new challenges to pure mathematicians.)

What physicists would like to find is a unique version of superstring theory that inevitably produces families of particles that can be identified with the known quarks and leptons. That may be decades away, if it ever comes, but the fact that the search may not be completely hopeless is borne out by several modest successes to date. For example, using one particular class of superstring models, Dimitri Nanopoulos, of the University of Wisconsin, and Keith Olive, of the University of Minnesota, found that some of the simplest versions of the theory lead naturally to a prediction that the electron neutrino should have a mass of about three one-millionths of an electron Volt, the mu neutrino should have a mass of 0.01 eV, and the tau neutrino should weigh in at 30 eV. Such a combination of masses would perfectly account for the dark matter in the Universe.

We don't want to get entangled in the details of superstrings here. But we cannot leave the topic without mentioning one interpretation of the equations that made headlines in the mid-1980s—the idea of "shadow matter."

Breaking Up Is Not
So Hard to Do

Symmetry is the key to the modern understanding of particles and forces. At very high energies, there is no distinction between, for example, electromagnetism and the weak force, and they are described as one force, the "electroweak." The simplest way to understand this is in terms of the masses of the particles that carry the forces. Electromagnetism is carried by photons, which have zero mass. The range of a photon is, in principle, infinite—the farthest quasar, at a redshift beyond 4.5, can exert an electromagnetic influence on the Earth itself (which is, indeed, what happens when a photon from a quasar strikes a photographic plate inside a terrestrial telescope). In practice, because positive and negative electrical charges are in balance, electromagnetic forces are not important across such very large distances.

The particles that carry the weak force, by comparison, have masses only a little less than one hundred times the mass of a proton. In order for one particle to influence another through the weak force, these force-carriers (known as bosons) have to be created. A particle with mass less than a proton cannot, obviously, "make" such a massive boson out of its own substance; the boson has to appear out of the vacuum, as allowed by quantum uncertainty, travel across to the neighbouring particle, and be absorbed back into the vacuum after making its presence felt. Because they are so massive, these virtual particles exist only for a very short time, and their range is limited to the distance they can travel in that time—roughly speaking, across the nucleus of an atom.

The electromagnetic force and the weak force become indistinguishable when there is so much energy around that these bosons can exist in profusion, in the same way that massless photons can be produced in profusion by stars today (or even by a modest electrical current flowing through the bulb of a flashlight). If the whole Universe was at a high enough temperature, weak bosons would become real, instead of virtual, particles. Such conditions existed in the Big Bang; when the temperature fell to the point where weak bosons could no longer exist as real particles, the symmetry between weak and electromagnetic forces broke.

Symmetry breaking is important not only because it explains how the complexity of the cold Universe we live in developed from the simplicity of the hot Big Bang, but because the changes associated with some forms of symmetry breaking could have provided the energy to push the Universe through a short-lived period of exponential expansion, the inflationary era, which smoothed out the wrinkles in spacetime and made it so flat. There is a lot more to symmetry breaking, however, than the distinction between electromagnetic and weak forces, or even the power of inflation.

The deepest symmetry we have mentioned so far is supersymmetry, the supposed symmetry between particles and forces that was broken very soon after the moment of creation. But some of the most promising versions of superstring theory contain exactly twice as much symmetry as this. There is "room" in superstring theory for just one more layer of symmetry, in which the *combined* world of particles and forces we know about is itself balanced by another, equally complex world of particles and forces that we do not know about. According to these theories, this would be the ultimate layer of symmetry, a splitting that occurred at the same time as gravity became distinct from the other forces of nature, just 10^{-43} seconds after the moment of creation.

Out of the Shadows?

When the Universe was very young and very hot, on this picture, there was a perfect symmetry in which all forces and particles were indistinguishable. Then, as gravity broke off from the other forces, the symmetry split into two smaller, initially identical symmetries. *One* of those smaller symmetries then went through successive further splittings, leading to the variety of forces and particles we know. What happened to the other?

Almost anything *could* have happened to it. As it split further in its own right, it could have produced a variety of particles and forces, possibly identical to the ones in "our" world, probably different. But the most important thing about that other symmetry—the other world—is that because it split from us at the time gravity became distinct, gravity is the only force—the only thing—that the two worlds have in common. We might be able to detect the other world by its gravitational influence on the matter of our world, but we could never interact with it in any other way.

What else could this other universe be called except the "shadow" world; and what else could its contents be called except shadow matter? You could be living at the bottom of an ocean of shadow matter, or walking through the base of a shadow mountain, and never know it. Science fact, it seems, has run headlong into the world of science fiction.

Shadow matter is an obvious candidate for the dark stuff of the Universe—a whole second universe, interpenetrating ours and expanding with it, sharing through the effect of gravity but otherwise undetected and undetectable. If the shadow world exactly mirrored our own, with the same amount of matter forming shadow quarks and shadow leptons (and, indeed, its own shadow dark matter, perhaps, in the form of shadow axions) then there could be shadow stars and planets within our own Galaxy. You may be reassured (or disappointed, depending on your taste for science fiction) to learn, however, that you are not living inside a shadow mountain. Although the two forms of matter could indeed interpenetrate to form a planet (or a double planet), calculations of the mass of the Earth and comparisons with the orbital motions of satellites show that there is less than 10 percent shadow matter inside our planet, and probably none at all. The prospect of shadow matter inside the Sun is even more slender—because this sort of dark stuff would sink to the core and exert a gravitational influence in the inner regions of the star without affecting it otherwise, it would make the Sun hotter in the middle, and that would show up in studies of neutrinos from the Sun (which, remember, actually suggest that the central part of the Sun is 10 percent *cooler* than standard theory predicts). The limit on the amount of shadow material inside the Sun is 0.1 percent, and the best guess is that there is none at all there. The clinching evidence against this kind of shadow matter comes from calculations of the way helium was manufactured in the Big Bang—the shadow matter would make the Universe expand too fast during the era of helium production, with more helium left over from the Big Bang than we actually see in old stars.

Very likely, the idea of a shadow world that exactly mirrors our own remains in the realm of science fiction. But there is no need to despair if you enjoy such speculations. Why, after all, should the shadow world have experienced exactly the same kind of symmetry breaking as our own world? Perhaps it contains different kinds of particles and forces, so that different rules of physics apply. A suitable choice of rules gets around the problem of helium production in the Big Bang and leaves the way open for speculators. All the material of the shadow world, for example, may decay into particles with zero mass. Or there might be a perfect balance between shadow matter and shadow antimatter, so that all matter in the shadow world annihilates into radiation. Or there might be one or more types of shadow particles that together contain just the right amount of mass to make the Universe (or universes) flat, and that stay spread out uniformly through space, never clumping together into stars and galaxies. And if that is too dull for you, imagine a shadow world in which the rules of physics are such that the stars are no bigger than a house here on Earth, so that a shadow star might fall on Stockton and the inhabitants would be none the wiser.

We are not, as you may have guessed, enthusiastic about shadow matter. There is too much room for speculation, and too little prospect of experimental or observational tests—and apart from anything else, there is no need for it. Particles that are known to exist (neutrinos) or that are required by our best theories (axions; miniholes) can perfectly well contribute all the dark matter, and even provide the critical density for a flat Universe, without leaving any room for shadow matter. Shadow matter is simply too much of a good thing, in gravitational terms. But we can never, by its very nature, prove that it does not exist. Cold dark matter particles might be detected in the lab or by their influence on the Sun and stars, and their properties studied; but you can never get a handle on shadow matter.* Cosmic strings, by comparison, cry out to be noticed. They almost certainly cannot, on their own,

*Unless you are lucky enough to find a minihole. Then, as Andrei Sakharov, of the P. N. Lebedev Institute, in Moscow, has pointed out, the Hawking evaporation of the hole will produce both matter *and* shadow matter. As a result, the hole will radiate energy more rapidly, and its temperature will rise faster, than Hawking's theory predicts. But first, catch your black hole.

provide all of the dark stuff that we require. But they can perhaps explain how the bright stuff got to be distributed in the way we see it today.

Strings and Things

The two main puzzles about galaxies are how individual galaxies form in the first place, and how and why they group together in chains, filaments, and sheets. Cosmic-string theory offers a possibility of answering both questions in one package. Our present theories of galaxy formation assume that the Universe used to be much smoother than we see it today, and that the lumpiness that galaxies represent grew out of some small initial irregularities, or seeds. By and large, cosmologists ignore the irregularities (except to use galaxies as convenient test particles to measure the expansion of the Universe) and deal only with equations that describe a smoothly expanding universe. But those very equations now provide a way to produce the required seeds naturally, out of the vacuum of spacetime. Symmetry breaking, so important to theories such as inflation and to modern understanding of the particle world, also provides three different, but related, kinds of flaw in the vacuum itself.

To a physicist today, the vacuum is very far from being the "nothing" that the term implies to the layperson. The vacuum out of which our Universe was born, perhaps through a vacuum fluctuation, contained a huge amount of energy, and possessed a high degree of symmetry, in the sense that there was no distinction between the fundamental particles and forces. The symmetry breaking that separated out those particles and forces was associated with a series of changes called phase transitions, in which the vacuum gave up its energy (helping to drive the expansion of the Universe).

This is rather like the way liquid water turns into ice. Compared with ice, liquid water contains a lot of energy. When water freezes, this energy is released, as latent heat; and the frozen water (ice) is less symmetric, because a crystalline lattice of ice (water) molecules does not look the same in all directions. The molecules in the lattice are aligned to make patterns, which we see in the beauty of a snowflake. A snowflake is definitely not the same in all directions.

The ice contains features that you never see in liquid water—boundaries between different crystalline regions that divide the ice up into smaller domains (for example, the boundary between one branch of a snowflake and the central body of ice from which it "grows"). Within each domain the ice may be relatively smooth, with the water molecules all pointing in the same direction; but the orientation of molecules in the crystal in one domain (one branch of the snowflake) will be different from the orientation of molecules in the domain next door.

Boundaries between different domains in a crystal (it need not be ice; any crystalline solid is as good an example) are usually like walls around the domains. But it is possible for other defects, as they are called, to form when a liquid crystallises. Some are point defects, where the molecules are aligned so that they seem to radiate outwards from a single point; others are one-dimensional lines. And all three types of defect can, in principle, occur in the vacuum of spacetime as a result of phase transitions and symmetry breaking when the Universe was young.

The kind of two-dimensional walls, or sheets, that are the most obvious example of this symmetry breaking in ice crystals are not observed in the Universe. A single domain wall that stretched across the visible Universe would contain much more mass (stored vacuum energy from the time before the phase transition) than all the matter we know about, including the dark stuff, and its gravitational influence would be obvious in the movement of galaxies. There *may* be domain walls farther away than we can see; but if so, they are being carried ever farther away by the expansion of the Universe.

At the other extreme, the kind of one-dimensional defects that form points in space turn out to be magnetic monopoles. Physicists were at first excited to discover that symmetry breaking in the early Universe could provide a means to manufacture monopoles, but were then embarrassed to find that those theories require the Universe to be swarming with monopoles that are not, in fact, detected. As we have mentioned, inflation provides a natural way to resolve this conflict; whatever its resolution, however, monopoles, like domain walls, have never been observed.

That leaves us with the intermediate type of defect, one-dimensional lines, or cosmic string, stretching across the Universe. Nobody has directly observed a cosmic string, either—but the existence of chains of galaxies in the Universe may be circumstantial evidence that cosmic strings exist.

Trapping the Vacuum

What, exactly, are cosmic strings? The concept stems from work by Tom Kibble, of the University of London, in the 1970s. It was taken up a few years later by Yakov Zel'dovich, in Moscow, and Alex Vilenkin, at Tufts University in America, each of whom realised its potential cosmic importance. They showed that during the symmetry breaking that occurred just 10^{-35} seconds after the moment of creation, some of the original vacuum state of the Universe could have been trapped inside linear imperfections in space. The best way to think of a piece of cosmic string is as a piece of the vacuum from that time, "frozen" and trapped inside a tube that has a diameter 10^{-14} that of an atomic nucleus. Because the string contains energetic vacuum from the birth of the Universe, it contains a great deal of mass (energy and mass are, after all, the same thing in this context). The actual mass depends on the exact time (and energy) when symmetry breaking occurred, but a favoured estimate would imply that each centimetre of cosmic string would contain 10 trillion tonnes of mass-energy; a piece of string a metre long could weigh as much as the Earth. You can see at once why the idea of cosmic string appealed to theorists who were trying to explain where the seeds that galaxies grow on came from. A loop of cosmic string with a diameter of a few hundred light-years could indeed help to hold a clump of gas together long enough, in the expanding Universe, for a galaxy to form. But there are complications.

Strings cannot have ends, for a start (this makes sense intuitively, since if there was an end, the energetic vacuum inside could leak out; it is also an inevitable consequence of the mathematics that describes strings). This means that either a string must stretch across the entire Universe (not just the bit we can see), or it must form a closed loop, like a rubber band. Cosmic string is like a (stretched) rubber band in another way—it has tension. The tension, like the mass of the string, is on a grand scale, and this sets any loop of string vibrating ("twanging") at high speed. These oscillations will occur as fast as they possibly can, almost at the speed of light, so that a loop of string one light-year in circumference will vibrate about once a year. Large amounts of mass-energy vibrating so rapidly must, according to general relativity, radiate away energy in the form of gravitational waves (more of this in chapter 8). Like a black hole evaporating away its mass through the Hawking process, but much, much more rapidly, a loop of vibrating cosmic string will lose energy and shrink away, eventually to nothing at all. This provides a very severe limit on the amount of mass that can still be in the form of cosmic string today. Such loops of string might have been important when the Universe was young and galaxies were forming, but they can form only a small part of the dark stuff needed to make the Universe flat today. If galaxies did indeed form around loops of string, then the galaxies we see today may be no more than the smile on the face of a vanishing Cheshire cat, showing where the body (the string) *used* to be.

This is particularly important because long pieces of string that stretch across the Universe will not be in straight lines. Instead, they will form a tangled mess, in which one string may cross over and tangle with other strings, or double back on itself and cross over another part of its own length. Ripples run up and down these strings at nearly the speed of light. Wherever strings cross, they will break and reconnect so that loops of string are split off and the original strings straighten out. Because the loops then radiate energy away, this ensures that string is never the dominant feature of the Universe. But it also leads to another very interesting phenomenon.

The way the Universe expands can be described in terms of the "Hubble length," a measure of the size of the Universe that is roughly the distance that light has had time to travel since the Big Bang. Since nothing can travel faster than light, objects that are more than this distance apart cannot interact with each other. That applies if the objects are separate galaxies, or if they are separate portions of an infinite length of cosmic string. The "wiggles" on a piece of infinite string are about as big as the Hubble length, so the loops that split off from the string are always about a Hubble length in diameter. This is true when the Universe is

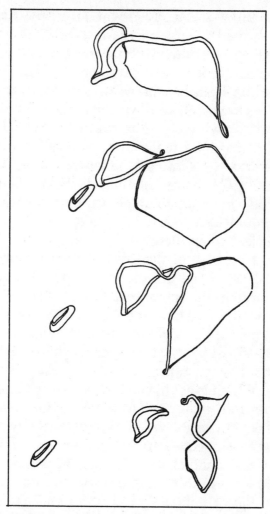

Figure 7.2 Four "snapshots" showing the computed behaviour of a loop of string that, during its oscillation, crosses itself and throws off smaller loops. (Courtesy of W. Press and R. Scherrer.)

small or when it is large; at any time, new loops of string are being broken off that can each be roughly as big as the observable Universe itself (smaller loops, of course, have already broken off, when the Universe was smaller). Each loop promptly begins to radiate gravitationally and lose energy, so at any time during the history of the Universe there must be a range of loops, graded in size from the Hubble length down to nothing at all. As new loops are always appearing at the biggest size possible in the expanding Universe, and small loops are always evaporating away, the result is that although the details may change, the overall nature of the pattern of large and small loops in the Universe stays more or less the same as the Universe expands—it is "self similar." This means that all the mathematicians have to do in order to calculate how the strings and loops should look today is to determine how the first loops looked when the Universe was young.

In round terms, for the first ten thousand years of its life the Universe, on this picture, contained only strings and hot radiation and spread-out matter. As the temperature fell, loops of string started to attract and hold on to clouds of gas and dark stuff. A galaxy could form around a small loop, while a larger loop would attract smaller loops (galaxies) to form a cluster; still longer strands of string could pull those clusters together in filaments and chains, and form new galaxies in a sheet-like wake. At every level, dark matter too would come under the influence of string. The statistical properties of the clusters and chains of loops of cosmic string that ought to form in this way resemble the statistics of the distribution of galaxies in clusters and chains in the Universe today. Once more, we are faced with a striking cosmic coincidence. We cannot prove that strings exist, but the similarity between the way string ought to be distributed and the pattern of galaxies on the sky is intriguing—the smile on the face of the Cheshire cat may, indeed, have been observed. So theorists have been encouraged to try to find out exactly how galaxies would grow around loops of string, and also to find ways of making more direct observations of the effects of string on the observable Universe today.

Making Galaxies

Cosmic string loops are a godsend to astrophysicists who try to account for the dark stuff primarily in terms of neutrinos. The problem about neutrinos is that they are "hot" particles that move very fast. In the early stages of the evolution of the Universe after the Big Bang, such hot dark matter homogenises by streaming through the baryonic gas and inhibits the growth of baryonic fluctuations. Galaxies can form in such a universe, but only after the hot dark matter has spread thin and begun to cool down—and that makes it very difficult to explain how galaxies as old as the ones we see around us can have formed in the time available since the Big Bang.

But loops of cosmic string cannot be blasted apart by fast-moving particles. They remain intact to act as gravitational seeds after the universe has expanded enough for the hot dark matter to be diluted and its influence weakened. At that point, baryonic matter can quickly begin to accumulate around the strings, producing structures that look very much like galaxies. Similar calculations can be carried through with a combination of strings and *cold* dark matter. Then, the opposite problem has to be tackled. *Without* strings, galaxies in a universe dominated by *hot* dark matter form too late; *with* strings, galaxies in a universe dominated by *cold* dark matter may form too soon.

The idea that massive string loops gather matter around themselves by gravity is the most obvious explanation of how they could act as seeds for the formation of galaxies. But there are other ways in which string loops can encourage the formation of galaxies, ideas that echo some earlier speculations about the foamy nature of the distribution of bright stuff across the Universe.

Ed Witten, of Princeton University, has suggested

that cosmic strings may act as superconductors. Any particles that happen to be trapped on the string will behave as if they have no mass, because the energy of the vacuum around them will be as great as the energy stored in their own mass—this is exactly equivalent to the way virtual bosons become real particles at high energies, unifying the electromagnetic and weak forces. Massless particles travel along the string without encountering any resistance, at the speed of light. If those particles happen to carry electrical charge, then enormous currents will flow unimpeded around the loops of cosmic string. When such a superconducting string oscillates, it radiates not only gravitational waves but electromagnetic waves as well, in copious quantities. A blast of electromagnetic radiation streaming out from a loop of cosmic string would push baryonic gas out of its way, forming an expanding bubble of material around the loop. Because the dark stuff does not carry electrical charge, however, it would be unaffected by the radiation and left behind. Galaxies would form where bubbles collide around the edges of voids full of dark stuff. We are left with a literally explosive scenario of galaxy formation.

Calculations made by Witten and his Princeton colleagues Jeremiah Ostriker and Christopher Thompson suggest that the resulting bubbles would produce a foamlike structure with filaments and sheets of galaxies surrounding voids up to 50 million light-years across, exactly as we see in the actual Universe. In that case, however, the string loops need not lie at the centres of galaxies, and the subtleties that distinguish galaxies formed against a background of hot dark matter from those formed against a background of cold dark matter would be lost.

This brings up another kind of cosmic coincidence. Some coincidences—mentioned in part 1 of this book, and discussed in detail in part 3—hint at the existence of Deep Truths, as John Schwarz puts it. They give us insight into special features of the laws of physics, which have to be the way they are if we are to be here at all to puzzle over them. Other coincidences are less deep. The way galaxies are distributed across the sky resembles the way cosmic strings, if they exist, *must* be distributed. But we can imagine other ways to make the pattern of galaxies. It doesn't *prove* strings exist, but it does encourage theorists to speculate further along these lines. Long strings should be moving through the Universe like cracking whips as wiggles move along infinite strings. These moving strings would leave "wakes" behind them, regions in which density had been increased, and in which galaxies might form. That, too, could explain why galaxies form in sheets separated by large amounts of seemingly empty space. There are many different ways in which cosmic string *could* explain the existence of galaxies in the Universe. So how might we expect to find strings, if they really do exist in our Universe?

Seeking Strings

Like all massive objects, strings affect nearby space-time through gravity. From far away—a distance much greater than the radius of the loop—a loop of cosmic string has a gravitational influence similar to that of any concentration of mass, such as a black hole. But close up, or when the radius of the loop is much bigger than the distance to the nearest part of the loop, another distortion of spacetime dominates.

Strings are not simply supermassive objects. They are cracks in the fabric of spacetime, defects in the structure of the vacuum. Space near a string has different properties from ordinary flat space, and this can be pictured by imagining an ideal, infinitely long, straight

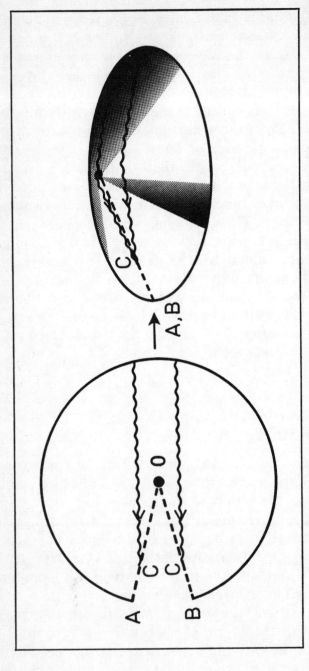

Figure 7.3 Illustration of space around a straight string. A small-angle wedge is removed from ordinary space, and the space becomes "conical" when the two lines $0A$ and $0B$ are brought together and identified with each other. Two light rays passing on either side of the string are focused at C.

string, lying still in flat space. The string distorts space around itself to make the *space* (not the string) conical. The way to get a handle on this is by considering a circle drawn around the line of the string. In flat space, the Euclidean geometry that we learned in school applies, and the ratio of the circumference of a circle to its diameter is *pi*, 3.14159. But if you draw a circle around a length of cosmic string and measure the ratio of the circumference to the diameter, you will find that it is a little less than *pi*. Or imagine travelling around a loop of cosmic string in a circle. In ordinary flat space, you would get back to your starting point after you had turned through 360 degrees. But if you were travelling around a piece of cosmic string, you would get back where you started before you had travelled around 360 degrees. It is as if a small angle had been cut out of space, and the edges pasted together to close up the gap.

The effect of this on matter is easy to visualise. Imagine two particles (or stars) moving through space parallel to each other. Because they are moving along parallel lines, like the lines of a railway track, they stay the same distance apart. But if the particles pass either side of a length of cosmic string, the distortion of conical spacetime makes their paths converge, so that eventually they will collide (this is the effect that compresses matter behind a moving string, and may encourage galaxies to form in the wake). The string distorts space as if the two particles were being pulled together by gravity—although this is not gravity in the usual sense of the term. It is a distortion of spacetime caused by the presence of a defect.

The speed with which the two particles begin to move together depends on how fast they go past the string—or, to put it another way, if we imagine the two particles at rest with a line of string passing between them, it depends on how fast the string is moving.

This makes it easy to answer the question that always arises when strings are mentioned. What would happen if one passed through the room in which you are sitting? The first point is that you would not be aware of the string through its mass in the usual gravitational sense. It is only when you view a closed loop of string from a long way away that it seems to have the gravitational field of a large mass. With a width less than that of a hydrogen atom, a string could cut at waist height right through your room, and your body, without your feeling a thing. But if the string were moving fast enough (perhaps at about the speed of light), the conical distortion of space behind it would quickly become apparent, as your head and feet (not to mention the ceiling and floor of the room) moved towards each other at a speed of several kilometres per second. It would be messy, and spectacular, proof that cosmic strings exist.

If the same thing happened to a star, the material making up the star would be squeezed, perhaps triggering a violent burst of nuclear reactions and making the star explode outwards. It is *possible* (we wouldn't go so far as to say likely) that an occasional stellar explosion could result from this string compression effect.

Conical space around a cosmic string would also affect the photons of the cosmic background radiation. Wherever a string moves transversely across the sky, as viewed from Earth, the radiation we see would be slightly cooler on the leading side and slightly hotter on the trailing side. If we ever find patches of the sky where the 3 K background radiation seems to differ from the average temperature, and especially if those patches have sharp edges, that might be taken as evidence in favour of cosmic string. A related effect of a length of string would be to bend light passing near it. If a piece of string happened to be passing between us and a distant galaxy, we might see two images of the galaxy, produced by light rays that had travelled along either side of the string and been bent towards the Earth. Massive objects, such as galaxies, also bend light rays that pass near them, producing multiple images in a similar way, and such a system is usually known as a gravitational lens. We will discuss the more familiar sort of gravitational lens in the next chapter; a key distinction between such lenses and the string effect is that gravitational lenses ought to produce odd numbers of images (three, five, and so on) whereas the string effect generally produces only two, which appear equally bright. So another test for the presence of strings is to search for regions of the sky where pairs of seemingly identical galaxies (or quasars) lie above and below a more or less straight line. There have, indeed, been claims that such pairs of images have been identified, but as yet none of those claims has stood up to closer scrutiny.

The more theorists investigate the possibilities afforded by string, the more fun it seems. The scenarios cannot all be right, but some of them might be. We have already mentioned the possibility of forming galaxies in flat sheets in the wake of a moving straight string; a small, fast-moving loop of string can do a similar job simply through its conventional gravitational influence on surrounding matter, pulling in mass behind itself to produce a tubelike wake. Both processes could have been operating back at a redshift of 200 or more, sowing the seeds for galaxy formation when the Universe was young. Loops of electrically conducting string could have started "blowing bubbles" at that epoch. High-energy radiation from these strings, in the X-ray or gamma-ray bands, emitted at a time long before the first galaxy formed, might be identifiable today. Different forms of string can produce both sheets and filaments of galaxies, even if the strings that did the job have long since moved far away from where we see the galaxies today, or have even evaporated completely.

Fast-moving loops that radiate energy asymmetrically will generally get faster, accelerating towards the speed of light, while their mass diminishes. On the other hand, if this rocket effect (due to either gravitational radiation or photons) acts to slow them down, loops that are born travelling at high speed could even slow to a halt, and then pick up speed in the opposite direction. While it is moving slowly, such a loop can gather mass around itself by gravity, and when it begins to speed up again it has to drag this mass along with it. If it accumulates enough mass, it cannot succeed in shifting it. Instead, trapped in the gravitational grip of the matter that was originally attracted by its own gravity, the loop may circle around in an orbit inside the accumulation of matter. If that matter forms a cluster of galaxies, we might look for traces of cosmic string in the form of an unusual, energetic galaxy displaced from the centre of such a system.

The observable consequences of strings depend on how heavy they are—on their mass per unit length. String theories relate this directly to one of the basic constants of unified theories, which is not yet pinned down by experiment. Were astronomers ever to find unambiguous evidence for gravitational lensing by a string, they could determine the fundamental mass quite straightforwardly. If strings indeed constitute the initial fluctuations from which galaxies formed, then we can already infer this mass to within a factor of 2. If theoretical physicists, using a different line of reasoning, were to come up with a similar mass as a requirement of their theories, this would suggest that strings did indeed trigger galaxy formation—otherwise, the agreement of the two estimates would be a simple coincidence. Moreover, there is a real prospect of de-

tecting the background of gravitational waves generated by strings, as we shall see in the next chapter.

Strings much lighter than those we have discussed
might also exist. If there are equally few of them as the
more massive strings, then they play no important role
in the evolution of the Universe. If, however, they do
not "reconnect" when they cross one another, splitting
off ever-smaller loops, then it is possible that this lighter form of string could form a tangled network that
has a total length so great that it contributes significantly to the dark matter. And that, perhaps, is a suitably awesome note on which to leave the subject of
cosmic string—except that it must, inevitably, crop up
once again as we turn our attention to ways in which to
probe the dark matter content of the Universe using
gravity's telescopes.

CHAPTER EIGHT

★

Gravity's
Telescopes

TWANGING COSMIC STRINGS are not the only things that make gravitational waves. General relativity describes gravity in terms of variations in the fabric of spacetime. It is a geometrical theory, all to do with curvature; the only odd thing about it, from an everyday perspective, is that what is being curved can be just empty space. But this is, perhaps, easier to accept today, when physicists talk about a vacuum seething with energy, and virtual particles popping in and out of existence, than it was when Einstein first introduced the concept. The way to remember how matter and space interact is through a simple couplet:

> Matter tells space how to curve
> Space tells matter how to move

A large mass like the Sun, as the couplet tells us, curves space in its vicinity. A smaller mass, like the Earth, follows the line of least resistance in that curved space.*

---

*The smaller mass also curves space, of course, and the line of least resistance is really determined by the combined curvature from both masses.

We see the effect as a force, gravity, pulling us towards the Sun, and holding our planet in orbit around the Sun. That orbit is the path of least resistance in curved space. But where do gravity waves come in?

Think of matter as solid lumps embedded in a stretched rubber sheet, spacetime; when one of those lumps vibrates, it sends ripples out through the sheet, and those ripples will set the other lumps of matter vibrating in sympathy. That is the principle behind gravitational radiation, and also behind the detectors with which physicists hope to measure gravity waves. The complications, such as they are, arise partly because space is actually three-dimensional, not a two-dimensional sheet, and, more importantly, because gravity waves are so very feeble that it pushes modern technology to the limit to hope to measure them. As far as the comparison is meaningful, since we are not really comparing like with like, gravitational radiation is only about 10^{-40} times as strong as electromagnetic radiation.

Gravitational waves are produced by moving masses, in a similar way to the production of electromagnetic waves by moving electrical charges. An isolated, perfect sphere of mass, however, does not radiate gravitational waves. The amount of radiation a mass can emit, according to Einstein's theory, depends on what is known as its quadrupole moment, a property related to its shape. An American football has a large quadrupole moment, but a soccer ball has none. Gravitational waves are in a form known as quadrupole radiation, and they have a distinctive effect on spacetime as they pass through it.

The best way to visualise what is going on is by thinking of a flexible circular ring. When a gravity wave passes it, the ring is stretched in one direction and squeezed in another, at right angles, simultaneously. It becomes an ellipse. Then, the pattern reverses, and what was the long axis is squeezed while what was the

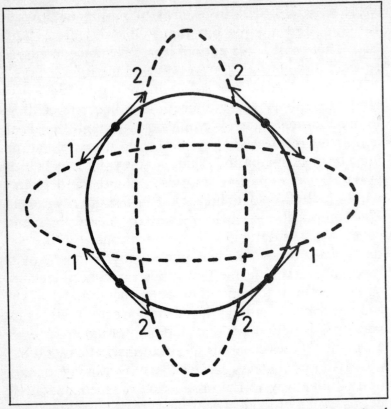

Figure 8.1 A circular ring is distorted into an ellipse when gravitational waves propagate through it. Waves could be detected by monitoring the position of the four masses, one in each quadrant (or, indeed, of any three of these, forming an L).

short axis is stretched. This pattern of alternate squeezing and stretching, in two directions at right angles to each other and out of step, is the characteristic "signature" of quadrupole gravitational radiation. It is not just the ring that is actually being stretched and squeezed, but the fabric of space itself. Four masses, placed one in each quadrant of the circle, will move rhythmically in and out, as though they are feeling a periodically varying gravitational (tidal) force. In fact, just three "test masses," marking an L shape, are enough to monitor the passage of gravity waves—*if* you have a sensitive enough means to measure the tiny movements that the passage of gravitational radiation causes. The effort will certainly be worthwhile if and when we can measure directly gravity waves from cosmic strings, supernova explosions, and other cosmic sources—perhaps shedding some light on dark matter in the process.

In a hypothetical universe where no electrical charges existed, the only kind of radiation would be gravitational. Electromagnetic radiation dominates in our Universe as a consequence of the 40-odd powers of 10 by which electrical forces overwhelm gravity on the short scale. But this doesn't mean that we can ignore gravitational waves entirely.

Making Waves

A long straight bar of material, spinning around like a two-bladed propellor, would be a good source of gravitational quadrupole radiation. Viewing the spinning bar sideways (in the same plane as it spins), you would see it first full length, across the line of sight, then pointing to you end on, and very narrow, then full length again, and so on. This is rather like the repeated squeezing and stretching of space by gravity waves, and the motion of the bar does indeed produce that kind of radiation. A spinning dumbbell, or two stars in orbit around each other, do similar jobs of producing radiation. A binary system in which the two stars were very close together, and orbiting very rapidly, would be the best source, and in one such system the effects of gravitational radiation have been discovered.

This is a system known as the "binary pulsar," although in fact it contains only one pulsar (a rapidly spinning neutron star that radiates at radio frequencies) orbiting closely around another neutron star that is not a radio source. Pulsars are a delight to astronomers because their pulses of radio noise (produced by a lighthouse effect as the pulsar spins) follow each other with exquisite precision; they are the most perfect clocks known, other than the vibrations within atoms that form the basis of modern scientific timekeeping on Earth, the atomic clocks (and some pulsars may even beat atomic clocks at timekeeping). Variations in the radio pulse rate from the binary pulsar, timed with microsecond accuracy, reveal its orbital motion around its companion. The apparent pulse rate speeds up when the pulsar moves towards us, and slows when it recedes— essentially a version of the Doppler effect. The period of the pulsar's orbit is very slowly decreasing. This means that the two neutron stars are moving slightly closer to each other as time goes by, which in turn means that the binary system is losing energy. General relativity tells us exactly how much gravitational radiation this system should be producing; it turns out that the predicted rate of gravitational radiation exactly matches the measured rate of loss of energy from the system. This is one of the greatest triumphs of Einstein's theory, and one that makes researchers confident that gravitational radiation will be measured directly, here on Earth, in the 1990s.

It was very nearly detected in 1987, when the supernova in the Large Magellanic Cloud exploded. When such a star dies, its core collapses suddenly inwards, and this collapse ought, relativity theory tells us, to produce a burst of gravity waves. The strength of the burst depends on how irregular and asymmetric the collapse is. A strictly spherical collapse radiates nothing. But even if the infall had been as chaotic as it could conceivably have been, the burst from supernova 1987A would still have been just one-tenth of the power needed to trigger existing detectors, by the time it arrived at Earth. As more sensitive detectors become operational, not just supernovae but vibrating or colliding black holes could also be detected (assuming they exist). We have already mentioned the potential of cosmic strings as gravity wave generators. The grandest speculation of this kind, however, is that the Universe might be filled with a background of gravitational radiation, left over from the violent events that occurred during the Big Bang and in the era of galaxy formation, comparable to the background of electromagnetic radiation. It might seem a daunting prospect to measure any of these effects, since they are calculated to produce a distortion of space on Earth amounting to only about one-millionth of the diameter of a proton for every metre of space being measured; but experimenters really do believe that they will have instruments that sensitive running very soon.

Measuring Waves

The experimental challenge was taken up by Joseph Weber, at the University of Maryland, in the 1960s. He built detectors based around large cylinders of aluminium, designed to vibrate as gravity waves passed by. Twenty years of research have failed to discover any such waves, almost certainly because they are too feeble to have a noticeable influence on these detectors.* But now a second generation of detectors, based on the same principles but a hundred thousand times more sensitive, are beginning to come into operation.

A typical "resonant bar" gravity wave detector is a cylinder of aluminium weighing 4,800 kilograms, cooled

---

*General relativity, in fact, *predicts* that any gravity waves likely to be around will be too feeble to be picked up by Weber's detectors. If Weber's detectors had found gravity waves, general relativity could not have explained such strong gravitational radiation in the Universe today—unless, by coincidence, their source were very close to the Earth, or (for instance) the galactic centre was unusually active at present.

by liquid helium to a temperature of 4 K (–269 degrees C). It has to be so cold in order to minimise the thermal vibrations of the atoms in the bar, and it has to be kept in a vacuum chamber to avoid being buffeted by molecules of air. A transducer attached to the bar converts any oscillation, such as the stretch and squeeze of a passing gravity wave, into an electrical signal, which is then amplified using superconducting technology. These amplifiers are so sensitive that they can indeed record vibrations in the bar corresponding to movements a thousand times smaller than the diameter of an atomic nucleus.

The main problem with this sensitivity is that any vibration, not just a gravity wave, will trigger the detectors. But advanced detectors of this kind are now becoming operational at Stanford and at the University of Maryland in the United States, in Rome, in Australia, and at other sites around the world. Researchers should be able to pick out genuine astronomical sources by their effect on each of these detectors in turn, and the time delay between the different detectors being triggered will give an indication of where the waves are coming from.

In another, quite separate approach to the search for gravity waves, other experimenters are working with beams of laser light. These experiments are very much like the idealised example of measuring changes in a circular ring. Large masses, with mirrors attached to them, are placed at two opposite corners of a square (it needn't actually be a square, but let's keep it simple) and laser beams are shone onto the mirrors from a third corner, the angle of the "L." The light from a single laser beam is split into two beams, and one sent out to each mirror and reflected back. When the two beams return, they are combined to produce an interference pattern of light. If the lengths of the two sides of the square change as a gravity wave passes, each laser beam will be affected, with one having farther to travel and the other less far to travel. The interference pattern will change as a result, revealing the passage of the gravity wave.

All of this involves running the laser beams through pipes evacuated to a very pure vacuum, about a metre in diameter and several kilometres long. Two such detectors are planned in the United States, to run in Southern California and in Maine; others may be built in Scotland and West Germany. Each will cost about the same as a large optical telescope, and if they work as planned, astronomers may one day be observing traces of gravitational radiation from supernovae in other galaxies and from other catastrophic events, such as collisions between two neutron stars in a binary system, or stars orbiting a black hole at the heart of the Milky Way.

Supernovae should yield gravitational wave pulses, but how strong these are depends crucially on details of the explosions and, in particular, on how symmetrical they are. Binary systems, on the other hand, are guaranteed to have big quadrupole moments, so even a pessimist could count on radiation from those. The only question is, can we detect it? About a hundred million years from now, gravitational radiation will have ground down the binary pulsar's orbit so small that the two neutron stars will revolve around each other hundreds of times per second, instead of once every eight hours (as they do today). The gravitational radiation emitted will then be enormously powerful, and during the final plunge, when the stars collide, coalesce, and form a black hole, up to 10 percent of their total mass-energy will be transformed into a burst of gravitational waves lasting only a few milliseconds. We do not know how many binary neutron stars of this type there are in the Galaxy. A reasonable guess would be a hundred or so. If each had a lifetime of a hundred million years, then one would "die" in this way every million years—such events are ten thousand times rarer than supernova explosions. A laser interferometer capable of detecting such a burst of gravitational radiation coming from several hundred million light-years away would, however, have more than a million galaxies like our own within range. The expected rate of detections would then be one a year—sufficient motivation for experimenters, who would be unhappy if a lifetime's labours yielded only null results (only a few experimenters derive sufficient satisfaction solely from meeting the technical challenge of devising sensitive equipment, regardless of whether it actually detects anything).

More powerful bursts still could come from the massive black holes that lurk in galactic centres. Mergers between pairs of galaxies are not uncommon. If there were a black hole in the heart of each such galaxy, the two holes would settle towards the centre of the merged system, forming a binary. The binary would emit gravitational waves, and eventually coalesce, releasing perhaps a hundred million times more energy than two coalescing neutron stars. The wavelength of this radiation, however, would be a hundred million times longer, because bigger objects are involved—the burst would last for hours, not milliseconds. Unfortunately, bars and laser interferometers on Earth are not sensitive to such slow waves, because of background vibrations caused by seismic activity, changes in the weather, and other terrestrial events.

Loops of cosmic string would emit powerful gravitational radiation of longer wavelength still—one cycle per year, or even slower. For these ultraslow waves, nature has provided a detector for us, in the form of single pulsars spinning rapidly with a precision far better than terrestrial clocks. Fast-spinning pulsars provide both the best evidence for gravity waves, from the binary pulsar, and the only constraint on how much *background* gravitational radiation there may be in the Universe. The fastest pulsars spin once every few milliseconds, producing precisely timed ticks of radio noise a few milliseconds apart. They are known, with a slight exaggeration, as *millisecond pulsars*. Atomic clocks have an accuracy of about one part in ten thousand billion (one in 10^{-13}). A pulsar can be even more accurate, "losing" less than a microsecond per century.* When the first millisecond pulsar was found, there was no way to test this, because there was nothing accurate enough to compare it with; but now several millisecond pulsars are known, and by comparing them with one another astronomers hope to be able to establish a timekeeping system, a cosmic clock, even more accurate than the atomic clocks. Any background of gravitational radiation that fills the Universe will distort space between us and the pulsar as the waves pass; the resulting jitter could affect the apparent regularity of the cosmic clock. This effect provides a sensitive probe for very-low-frequency gravitational waves—ripples in spacetime with wavelengths of a few light-years. Since no such effect has yet been seen, we can say with confidence that the amount of mass-energy stored in gravitational radiation of this kind is no more than one-millionth of the amount needed to make the Universe flat by this means alone. This upper limit is already very interesting to string theorists, since it is close to the level of the expected gravitational wave background from string loops (see chapter 7). When the

---

*They actually slow down more rapidly than this, but in a steady and predictable manner. What matters, if we are using such a system as a clock, is the amount by which the clock might deviate from this steady change, and that could be a fraction of a microsecond per century.

pulsar timing data have been gathered for a few more years, the results will be still more restrictive. If a wave background is still not detected, we shall have to conclude that strings do not exist (or, if they do, that their mass is too low to have triggered galaxy formation).

Gravitational radiation by no means dominates the Universe, but its eagerly anticipated discovery will provide astronomers with a new kind of telescope with which to probe energetic objects. Those gravitational telescopes are sure to reveal previously unsuspected features of the Universe, as well as help us to get a handle on dark matter and, perhaps, strings. Meanwhile, dark matter itself may be bending spacetime in its vicinity sufficiently to produce gravitational lenses that give us a view of objects so distant that they could never be seen without this other form of gravitational telescope. It may, indeed, be making a spectacle of itself, even while it remains unseen; gravitational lensing provides the nearest we can ever come to actually seeing the dark stuff that dominates the Universe.

Gravitational Lenses

Light bending is just about the most familiar and well-tried feature of general relativity. The theory first appeared in print in 1916, complete with Einstein's prediction that light would follow curved paths through space distorted by the presence of matter. In 1919, this light-bending effect was measured during an eclipse of the Sun, showing up as a displacement in the positions of the images of stars that lay near the Sun on the sky (but far beyond in space) at the time of the eclipse. Light from those stars, coming from behind the Sun, had indeed been deflected, by exactly the amount Einstein had predicted, as it passed the edge of the Sun. The gravitational bending of light was first seen, and photographed, more than seventy years ago. This is the basis of gravitational lensing.

A sufficiently massive object lying between us and a distant star would bend light so much that it might produce two images of the distant star, as viewed from Earth. In 1936, Einstein himself investigated this possibility, and proved that if a compact, massive object can indeed, under the right circumstances, create two separate images, one (and sometimes both) being magnified. If the alignment were perfect, the stars would appear as a complete ring of light encircling the "lens."

Things become a little more complicated, and even more interesting, if either the lens itself or the object that is being lensed is an extended object, such as a galaxy. If a black hole with a hundred times the mass of our entire Galaxy lay halfway between us and a distant galaxy (not, we must admit, a very likely possibility), the image of that galaxy on the sky would be in the form of a bright ring of light, from the part of the galaxy exactly behind the black hole, with two images like the ones Einstein described, one bright and one dim, on opposite sides of the ring. With a much less massive (and correspondingly more plausible) black hole doing the imaging, the ring would be too faint to see, and the two images of the distant galaxy (or quasar) would appear on their own. In 1979, astronomers found two quasars whose images are just six seconds of arc apart on the sky—this is about the angle covered by a tennis ball at a distance of five kilometres. These two quasars are so alike, in terms of their colour and redshift, in particular, that they were soon regarded as the first identification of a pair of images from a gravitational lens; we now know that there is a large cluster of galaxies, which includes a giant elliptical, in just the right place to be acting as the lens that is producing the two images of one quasar.

More than half a dozen gravitationally lensed systems are now known. The exact number you choose depends on the date (observers seem to find about one more system of this kind each year) and your gullibility (sometimes, enthusiasts claim that pairs of quasars are gravitational images, but later investigation shows that they really are two distinct quasars). But there is another very interesting system, apart from the initial discovery, where the object that causes the lensing has also been identified—in this case, as a large disc galaxy relatively near to our own Galaxy. This system is particularly interesting because it shows what happens when the object that makes the lens is large and definitely not spherical. The first observations suggest that the light from a distant quasar has been split and bent to form *three* images, a tiny triangle on the sky with sides a few seconds of arc long. More sensitive measurements later revealed a fourth image, completing a square surrounding the galaxy's central regions; and there may even be a fifth almost exactly underlying the centre.

The number of images is crucial for any attempt to determine the nature of dark matter in the Universe using gravitational lenses. If the lens is a black hole, it should produce two, and only two, images. If the lens is a spread-out object such as a galaxy, it should produce at least three images, perhaps more, but definitely an odd number.* And if the "lens" is actually caused by the distortion of spacetime behind a cosmic string, there should be two equally bright images. Over the past couple of years, however, astronomers have been even more excited about the discovery of another kind of arc across the sky.

*It is at least *slightly* embarrassing that observers have yet to find a third image in the first gravitationally lensed system identified, because the lens there does seem to be an extended object. There are ways around this; perhaps the third image is too faint to be seen, perhaps the lens is actually a black hole in the intervening cluster of galaxies, and so on. The study of gravitational lenses in the real Universe (as opposed to in the mathematical dreams of theorists) dates only from 1979 and still has loose ends to be tied up.

Luminous Arcs

These features of the Universe are *big*. They stretch around almost perfectly circular arcs, bits of circles, for a length of more than 300,000 light-years; the width of each arc is about 30,000 light-years. Two of these huge, almost perfect arcs were discovered in the mid-1980s; each seems to lie in a cluster of galaxies. They each stretch across a distance three times bigger than the size of our Milky Way Galaxy, and are the largest known continuous bright objects in the Universe; although they look very small, it is because they are so far away. A third, more wispy arc was discovered at about the same time.

The discoveries were reported almost simultaneously by French astronomer Bernard Fort and colleagues at the Toulouse Observatory, and by Roger Lynds and Vahe Petrosian in the United States. The French team almost immediately guessed that the features might be produced by the gravitational lens effect—that they might, indeed, be "Einstein rings." At first this suggestion met with little response. Lynds and his colleagues initially suggested that the arcs might be expanding shells of material, blasting outwards from some cosmic explosion, perhaps caused by a collision between galaxies, and perhaps a common feature of clusters of galaxies. The theorists had a field day for a while, coming up with all kinds of exotic ideas to explain how such perfectly circular arrangements of stars could form. But a splash of observational cold water soon came from the French team, which showed that different segments of a single arc all had exactly the same spectrum and must therefore all be part of the same structure. Proof of the nature of the arcs came when the redshift of those spectral features was measured.

In the best example of this newly discovered astronomical phenomenon, the arc looks as if it is part of a cluster of galaxies known as Abell 370. But this cluster is at a redshift of 0.374, while the light from the arc has a redshift of 0.724. The light from the arc actually originates almost twice as far away in the expanding Universe as the distance to the cluster. It is a magnified and distorted image of another galaxy. Intriguingly, though, calculations show that the clusters cannot produce powerful enough lensing unless they contain at least ten times as much mass as we can see in the form of bright stars in the constituent galaxies. This exactly matches the broad picture in which 90 percent, or more, of the gravitating stuff is dark.

There are many intriguing implications. First, the spectrum of light from the arcs seems to match the spectra of disc galaxies (the averaged-out light of billions of stars), magnified and brightened up to twenty-five times. This means that they may tell us what ordinary galaxies, as opposed to quasars or radio galaxies, were like when the Universe was less than half of its present age. A lot more analysis needs to be done, but it is already clear that the light from these galaxies contains a lot of ultraviolet radiation, typical of hot young stars—exactly what you would expect if we are viewing galaxies in the early stages of star formation. In addition, although these two large arcs were noticed simply because they are so large, the alignment needed to produce them must be a very rare occurrence. There should be many more systems in which there is a less perfect alignment, and only fragments of the Einstein ring are produced—and some of the peculiar objects that astronomers photograph in clusters of galaxies may well also be fragmentary images of very, very distant galaxies. Gravity really does provide a telescope with which we can view things as they were long ago and far away; but you don't necessarily need a whole cluster of galaxies to make an effective lens to use in gravity's telescope.

Shedding Light on
Dark Matter

Light from the most distant objects in the Universe, the quasars, may indeed be revealing the nature of the matter that holds our Galaxy together. The double images on the sky formed by light from some quasars may be produced by galaxies, 90 percent of whose mass is in a dark halo. Fine structure, details within those images, may, furthermore, tell us whether these haloes are made of very massive objects (VMOs) or brown dwarfs (Jupiters), and thereby decide between these two rival baryonic candidates for the halo dark matter.

If VMOs provide a substantial fraction of the halo dark matter, there could be a million black holes, each with the mass of a million Suns, together providing ten times more mass than all the bright stars of the Milky Way system put together. The probability of seeing the lensing effect due to an object in the halo of our own Galaxy is only about one in a million. But it is much more likely that light from a very distant quasar will be lensed, on its way to us, by an object in the halo of another galaxy roughly halfway along the line of sight. If the light passes through the haloes of several galaxies, as it must for very distant quasars, then clearly there is a still bigger chance of it being lensed. We are *not* talking now about the lensing effect of a whole galaxy, or of a cluster of galaxies (what you might call "macro" lensing); instead, we are considering light from a distant quasar being bent by a *single* star- or planet-sized object in the halo of a galaxy between us and the quasar (microlensing).

Such an event may be fairly likely, but would it be observable? Surprisingly, the answer is *yes*. If the microlensing is caused by a VMO of about a million solar masses, and this object is about halfway out across the visible Universe (at half the Hubble distance), then

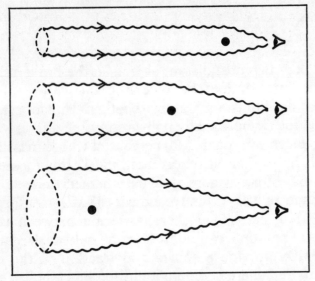

Figure 8.2 The focussing of light from a very distant source by a compact mass. The lens has, in effect, a short focal length for light passing close to the compact object, and a longer focal length for more distant passages. A given object is therefore more efficient (i.e., has a larger cross-section for forming images) when it acts as a lens with very long focal length. For this reason, a black hole in a remote galaxy is more likely to cause detectable lensing than a similar hole close at hand in our own galactic halo.

the two images it produces will be separated by an angle of one-thousandth of a second of arc, as viewed from our Galaxy. This is several thousand times less than the separation of the widest known pairs of images in the gravitationally lensed systems we have discussed so far, and no *optical* images could be sharp enough to reveal such fine structure. But by linking together electronically *radio* telescopes on opposite sides of the world, astronomers can create the effect of a single instrument with the same resolving power as a radio dish the size of the Earth. This technique, known as interferometry, could measure such a small separation between the components of a double source. Many candidate objects have been observed, with no clear

case of double images. The statistics suggest that microlensing is so rare that no more than one-tenth of the
matter needed to make the Universe flat can be in the
form of VMOs, which each have a mass of a million
Suns (which, of course, gives at least a gentle push to
close the door on the possibility that most of the halo
mass in our own Galaxy is in the form of VMOs).

What about brown dwarfs or Jupiters? In that case,
the same effect would produce images with a separation, viewed from Earth, of less than a *millionth* of a
second of arc. This is far too small to be measured. But,
equally, it is such a small angle that an object half the
Hubble distance away from us and travelling at the
modest speed (by astronomical standards) of a hundred
kilometres a second will cross an angle of a millionth of
a second of arc in a few years. This could provide a way
to make images of distant quasars twinkle quickly
enough to be noticed in a human lifetime. The situation
is complicated if the quasar light passes through several galactic haloes and has the chance to be lensed
several times on its way to us, and also because other
effects, intrinsic to quasars, cause them to vary. Once
again, however, such limited observational evidence as
we have so far seems to rule out the possibility that
more than one-tenth of the matter needed to flatten the
Universe could be in this form.

There are at least some hints from these microlensing
studies that haloes are neither predominantly VMOs
nor Jupiters, although some such objects may be present.
That suggests that the halo stuff itself is nonbaryonic,
spread out, dark matter. Which raises another question: If a typical galaxy is 90 percent, or perhaps 99
percent, spread-out dark matter, with the visible stars
merely a kind of icing on the cake, is it possible that
some of the cakes have been left uniced? Are there, in
fact, dark "haloes" that do not contain bright galaxies

at all? Gravitational lensing may provide the means to answer that question, as soon as the Hubble Space Telescope gets into orbit.

Dark Galaxies

Recent work on galaxy formation, such as the computer models involving cold dark matter, suggests that there may well be dark haloes in which galaxies have failed to form. Because light gets bent by matter regardless of whether that matter can shine (as the example of a black hole makes clear), these "invisible" haloes should reveal their presence by distorting light from distant bright objects. One of the most puzzling features of the known handful of cases of gravitational lensing in the Universe is the large separation between the two images in some examples. This is much bigger than you would expect if the object doing the lensing was a single ordinary galaxy or a black hole. Another puzzle is, as we have mentioned, the fact that in some of these systems there is no trace of a bright galaxy (or galaxies) in the right place to produce the lensing effect. Both these puzzles arise with full force in the case of the system with the widest separation of images, a full 7.3 seconds of arc. Could this, and other such systems, be a result of lensing by dark haloes along the line of sight?

When the appropriate calculations are carried through, it turns out that a single extended halo is unlikely to form multiple images of a bright object that lies behind it, as viewed from Earth. The light is distorted, but not strongly focussed enough to yield multiple images. On the other hand, if *two* dark halo-type objects lie along the line of sight to a distant source (rather like the two lenses in a simple telescope) then multiple imaging *can* occur. The situation is quite complicated, involving lensing by two haloes at different redshifts, which may not be perfectly aligned with each other or with the distant quasar. In CDM models there may be more of these "failed" galaxies than there are visible, starry galaxies. In a flat universe the effect is most likely for dark haloes that lie at redshifts between 0.3 and 0.6, and then double lensing by such haloes ought to produce images separated by between 5 and 7.5 seconds of arc as viewed from our Galaxy. This is exactly in the range where the puzzlingly widely separated images are seen. With no intervening lens galaxy detected in careful searches of the sky, we feel that there is a strong case that one or two of the known gravitationally lensed systems involve lensing by two dark haloes.

Failed galaxies may be dark because the baryons in the halo have all condensed into dim objects, such as brown-dwarf stars. Alternatively, they might be haloes that contain only nonbaryonic matter, from which all the baryons have, for some reason, been pushed out. But perhaps the best bet is that they are a mixture of baryonic and nonbaryonic material, in which the baryons are spread out through the halo in the form of clouds of hydrogen gas (themselves laced with 25 percent helium produced in the Big Bang). In this section of our book, we have concentrated so far on the dark stuff that dominates the Universe, the nonbaryonic matter. But we should not forget that there may be a great deal of dark *baryonic* material as well. The galaxies we see, even if each contains ten times as much mass in dark stuff (possibly nonbaryonic stuff) as in the bright stars, still contribute an average density of only 10 to 20 percent of the amount required if the Universe is indeed flat. So most of the dark stuff, and most of the hydrogen atoms as well, may lie between the galaxies and clusters. There are a lot of unseen hydrogen atoms out there somewhere.

Some of the theories of galaxy formation that we have discussed suggest that the voids between bright galaxies are not really empty but contain large numbers of failed galaxies. And now, as we shall see in the next chapter, there is a way to investigate those failed galaxies by looking not at the way they bend light from distant objects but at the way they leave their imprint in the form of dark lines in the spectra of quasars.

CHAPTER NINE

---★---

The Lyman Forest: Emergence and Evolution of Galaxies

ASTRONOMERS MEASURE DISTANCES to galaxies and quasars in terms of redshift, the displacement of lines in the spectra of these objects compared with the wavelengths at which these lines occur in the laboratory here on Earth. These spectral lines are produced when electrons in atoms move from one energy level to another. The best way to think of these energy levels are as steps on a staircase. An electron can "sit" on any step, but there is no stable position for it to rest between steps. If the atom absorbs precisely the right amount of energy, an electron can jump up one, or two, or some other whole number of steps. But it can only jump a whole number of steps, because there is nowhere else to jump to. Then, a little later, it may fall back down, again by one, or two, or some other whole number of steps. The amount of energy involved depends on the size of each step, and the number of steps the electron jumps in one hop. Electromagnetic radiation, such as light, carries energy; the shorter the wavelength of the light, the more energy it carries. If the atom absorbs energy from

223

background light, this leaves a sharply defined dark line in the spectrum, where the light has been taken away. If the electron falls down the steps, it radiates energy, again at a very sharply defined wavelength, producing a bright line in the spectrum.

These lines can be measured in the laboratory. Their wavelengths can also be predicted, using quantum theory. Indeed, the success of quantum theory in explaining the spectrum of hydrogen was one of the great triumphs of physics in the early twentieth century.

Because hydrogen is the simplest element, involving just one electron in orbit around a single proton, it has the simplest spectrum and the easiest to calculate. The energy levels in hydrogen are very precisely known. This is particularly useful for astronomers, because hydrogen is also by far the most ubiquitous element in the Universe, making up 75 percent of all the baryonic matter, both in bright stars and dark clouds. A thorough knowledge of the spectrum of hydrogen, together with the redshift effect, is all you need to measure cosmological distances.

Even the spectrum of hydrogen contains many lines. Think of our staircase as having just six steps from top to bottom, with the lowest step representing the lowest energy level, closest to the proton that forms the nucleus of the atom. An electron jumping from step six to step one will produce a characteristic line in the spectrum. Any other electron, making the equivalent jump in another atom, will produce the same wavelength of radiation, adding to the strength of that line in the spectrum of a cloud of hot hydrogen gas. But this is only the beginning. An electron jumping from step five to step one will produce another, different line, as will a jump from step four to step one, and so on. The result is that the spectrum of a cloud of hydrogen contains many lines, one associated with every possible jump. All the lines in this set involve jumps that end on step one, and they will have a family resemblance. Then there will be another series of lines involving jumps that end on step two, a series involving jumps that end on step three, and all the rest (and we chose to consider only six steps; in reality there are more!). Just one of these sets of lines, however, is important to our present story.

The jumps that end on step one, for a hydrogen atom, were studied by Theodore Lyman, an American physicist, in the first two decades of this century. These lines are all in the ultraviolet part of the spectrum and involve higher energies than lines that appear in the visible part of the spectrum, at longer wavelengths than the UV. In Lyman's honour, this set of lines is called the "Lyman series," and the brightest line in the series is known as Lyman-alpha. Lyman-alpha occurs at 122 nanometres (nm) wavelength; this wavelength is both measured in the lab and predicted by quantum theory. Because the ozone layer of the Earth's atmosphere shields us from ultraviolet radiation, the Lyman-alpha line cannot be seen from Earth in the spectrum of the Sun or the spectra of other stars in our Galaxy. But Lyman predicted that the line ought to be present in the Sun's light, and when rockets lifted ultraviolet detectors above the stratosphere in 1959 (five years after Lyman died) his prediction was confirmed. Even in 1959, however, astronomers had no inkling of how important the Lyman-alpha line was to be in their work.

Quasars and Lyman-alpha

Quasars were first identified in the early 1960s, not long after the first direct observations of Lyman-alpha in sunlight. There are many spectral lines that can be used to measure the redshift of a quasar, but hydrogen

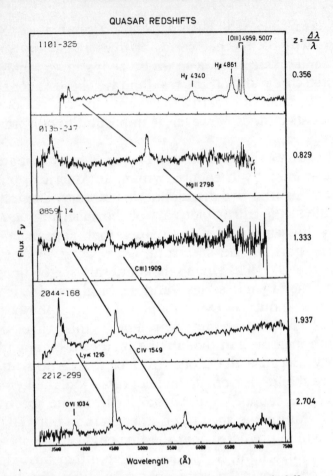

Figure 9.1 The spectra of several quasars with different redshifts *z*, showing how the optical spectra of high-z objects reveal emission lines that are normally in the far ultraviolet.

lines are best because hydrogen is so common. Because of the redshift, the light by which we actually see (or rather, photograph) a quasar is not at the wavelength at which it was emitted. The energy that reaches Earth in the visible part of the spectrum originates from shorter wavelengths—in the ultraviolet. It happens that very hot, energetic objects such as quasars radiate a lot of energy in the ultraviolet, and because of the redshift this energy may show up, to us, as a bright contribution to the blue end of the spectrum. It sounds paradoxical, but despite the *red*shift a quasar can actually look very *blue*, to human eyes, because the blue light we see used to be even "bluer"—that is, ultraviolet (a bigger redshift still, will of course shift the light all the way to the red end of the spectrum, so that the quasar actually looks red). With a lot of energy being radiated in the ultraviolet, the Lyman-alpha line of a quasar ought to be very strong. And with a big enough redshift, that line ought to be moved into the visible part of the spectrum, where radiation passes unaffected through the ozone layer of the stratosphere. In other words, it is actually possible to detect Lyman-alpha radiation from a high redshift quasar *on the ground*, without sending instruments into space. Lyman-alpha lines redshifted by as little as 1.7 (but no less) can just be detected, at a wavelength of 330 nm, using instruments on the ground. The Lyman-alpha emission line is generally so strong and clear that studies of its shape and energy can be used to infer details of the energetic processes that make the quasar shine so brightly. These are among the studies that suggest that quasars are powered by supermassive black holes. But this is far from being the end of the story. In a typical quasar spectrum, the Lyman-alpha line stands out as a high peak, like a very tall mountain. At wavelengths on the blue side of this line, however (at slightly smaller redshifts), there are very many weaker, dark lines, like a series of very narrow but steep valleys dipping down below the plain from which the Lyman-alpha mountain rises. These lines cannot be formed in the quasar itself—they are spread over a range of redshifts corresponding to distances of hundreds or thousands of millions of light-years in the expanding Universe. They must be caused

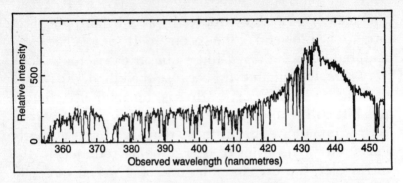

Figure 9.2 A spectrum of the quasar Q2206–199, with redshift 2.56, showing the many narrow lines that make up the Lyman forest. The broad Lyman-alpha *emission* line is on the right. (Taken by A. Boksenberg and W. Sargent, using the 5-metre telescope at Mount Palomar.)

when light from the quasar is absorbed by clouds of cold gas that lie between us and the quasar.

This forest of dark absorption lines was first noticed in 1971, but in the early 1970s spectroscopic techniques were not good enough to reveal many details of the forest. As spectroscopy has developed, however, astronomers have realised that they are looking at a whole series of Lyman-alpha lines, each one redshifted by a different amount. Like a distant searchlight silhouetting a nearby tree, the quasar light highlights hydrogen in clouds between us and the quasar. By the early 1980s, techniques had advanced sufficiently for an analysis of the Lyman forest to begin to yield information about those clouds.

Into the Forest

The study of the Lyman forest in the 1980s has been particularly exciting because it provides information *only* about regions of the Universe beyond a redshift of 1.7. Ordinary galaxies can be studied in detail out to a redshift of only about 0.3. This covers only a few billion

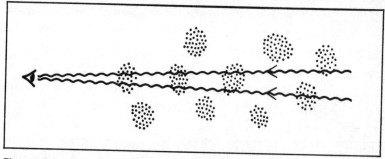

Figure 9.3 If the Lyman forests along two lines of sight—whether to neighbouring quasars or separate gravitationally lensed images of a single object—show some (but not all) lines in common, then the cloud sizes must be comparable with the transverse separation, as depicted here.

years of history, and a few billion light-years of space. All of the descriptions of bubbles, voids, sheets, and filaments of galaxies in the Universe are based on this relatively small portion of spacetime. The Lyman forest tells us about the early days, when the Universe was young—from about a billion years after the Big Bang to 4 billion years after the moment of creation. And what it tells us confirms the broad picture of a Universe dominated by dark matter, while eliminating some of the detailed models from contention and strengthening the credibility of the cold dark matter cosmology.

The light from a single quasar may contain dozens of Lyman-alpha absorption lines at different redshifts. By studying the details of a single line, astronomers deduce something about the conditions inside the cloud of gas that absorbed that particular wavelength of light.

Studies of the Lyman forest also show how big these clouds are. In some cases, very similar patterns of Lyman-alpha lines are seen in the forest from each of two nearby quasars. These may be different quasars that happen to lie nearly along the same line of sight, or they may be two images of a lensed quasar. What matters is that the light coming to us from two very slightly different paths contains the same Lyman forest features. This shows that some of the intervening clouds are big enough to cover both of the quasar images on the sky. In other cases, pairs of nearby quasar images have quite different Lyman forest features—their light has *not* passed through the same clouds en route to us. A typical dark cloud turns out to be about as big as a small galaxy, 35,000 light-years across. And that in turn, adding in the estimates of the number of hydrogen atoms and protons in each cubic metre of space, tells us that the mass of a typical dark cloud is between 10 million and 100 million times the mass of our Sun, very much in the range of dwarf galaxies like the Magellanic Clouds, but a lot less than the mass of our own Galaxy. Using quasar light, we can actually weigh and measure dark clouds in space perhaps 10 billion light-years away from the Milky Way.

Another important feature of the improved spectroscopic studies is that they show no traces of lines corresponding to anything except hydrogen. The clouds ought to contain 25 percent helium, if our calculations of the Big Bang are correct, but the appropriate helium lines are too far in the ultraviolet to be seen from the ground even at these high redshifts. When the space telescope flies, it should detect these lines, as the helium forest, and tell us whether there is indeed 25 percent helium present in the clouds. The fact that no lines of heavier elements are seen is confirmation that only hydrogen and helium emerged from the Big Bang— heavier elements are made in stars, and stars have not yet formed in these clouds.

Modern techniques tell us how big individual dark clouds are and what they are made of. We can also use them to give us another handle on the nature of the nonbaryonic stuff of the Universe, by looking at what holds an individual dark cloud together, and at how the clouds are distributed across the Universe.

Large-scale Lessons

If clouds like this existed in isolation in space, they would disperse fairly quickly, as astronomical timescales go. They would, in a sense, evaporate. So what *does* hold them together? One school of thought holds that they are embedded in still bigger clouds of much hotter material. In those larger clouds, there would be no neutral hydrogen, only free protons and electrons, and so no Lyman lines to be seen. The pressure of the hotter gas outside would, however, stop the smaller, cooler clouds that we see from expanding and evaporating away. If this picture were correct, then as the Universe expanded this confining cloud of hot material would have cooled, so that the confinement of individual dark clouds was eased and they could disperse, leaving only wispy traces of gas behind. Since there do seem to be more lines in the Lyman forest at higher redshifts, suggesting that the dark clouds have gone somewhere as the Universe aged, this possibility has to be taken seriously. But we prefer an alternative scenario, which relates the presence of these dark clouds very neatly to the cold dark matter cosmology.

In the cold dark matter scenario, galaxies form from the bottom up, as small clumps of matter group together to make bigger clumps, and so on. Early in the history of the Universe, there should have been many irregularities in the distribution of the CDM on a scale smaller than the galaxies we see today. These irregularities, the gravitational potholes, would trap baryonic matter, but there is no reason to expect that in all cases all of this baryonic matter would condense into stars. A deep pothole with steep sides would pull baryons into its centre, where stars would form; a very shallow pothole might not be able to trap baryonic gas at all; but in an intermediate range of possibilities, moderately deep potholes with moderately sloping sides would trap gas that stayed free to move about within the pothole instead of being concentrated into stars. The measured properties of the gas in the clouds that produce the Lyman forest exactly matches the calculations of the properties of this kind of pothole. The gas causing quasar absorption lines may indeed be confined in potential wells where something else, other than baryons, provides the main gravitational influence. The situation is very reminiscent of the measurements of the rotation of disc galaxies, which show that the galaxies would fly apart if they were not held in the gravitational grip of dark matter. The dark clouds of the Lyman forest will also fly apart if they are not being held together by something we cannot see. The simplest and most natural assumption is that the same kind of something—cold dark matter—is doing the holding together in each case.

But the Lyman clouds are different from the bright galaxies we observe in one important way. They are not distributed in a foamy structure surrounding empty voids. The redshifts of the Lyman-alpha lines in the forest seem to be distributed at random, apart from the tendency for there to be more of them at higher redshifts. If there are any voids that contain no Lyman clouds, then they must represent less than 5 percent of the volume of the Universe at the times corresponding to redshifts between 1.7 and 4.

This is a major discovery, which tells us that not only are bright galaxies poor indicators of where most of the mass of the Universe is, they are not even a good guide to where all the baryons are. And it fits beautifully with the idea of biased galaxy formation. Computer models tell us that in a Universe dominated by cold dark matter, bright galaxies should be clustered. But those studies also tell us that the voids should contain a density of dark matter nearly as great as the density in the surrounding sheets or filaments, and that there should be the usual mixture of a few percent baryonic matter mixed in with the CDM in the voids, some of it confined in "minihaloes." The Lyman-alpha forest could have been predicted on the basis of the CDM theory— had it not been discovered more than ten years "too early." However, this order of events reflects on the tardiness of theorists, not on the credibility of the theory itself. The Lyman-alpha forest is strong corroboration of the *simplest* version of CDM cosmology, in which the Universe is indeed flat. It tells us clearly that all parts of the Universe are pervaded by Dark matter *and* by baryons; the huge apparent voids are not empty, but just deficient in bright galaxies. And this is still not the end of the evidence about the nature of the Universe that can be gleaned from quasar light.

Heavy Signs of a Galaxy Wall

A typical quasar spectrum contains over a hundred Lyman-alpha lines, each at a different redshift. The Lyman clouds themselves do not contain heavy elements. But the quasar spectra do show a handful of lines belonging to heavy elements. These lines must be produced when light from the quasar passes through a galaxy like the ones we see in our neighbourhood. As stars have evolved in those galaxies, some have become supernovae and spread heavy elements, built up in their interiors, into interstellar space. So hydrogen clouds in a galaxy like our own are laced with heavy elements, and light from a quasar passing through a cloud of hydrogen in a galaxy like our own will be stamped with the imprint of those heavy atoms. Bright galaxies, however, are much less common than the Lyman clouds. The chance that the line of sight to a quasar will pass right through a bright galaxy is correspondingly modest. In fact, almost every quasar spectrum shows a few lines due to heavy elements. This can only mean that all galaxies are embedded in large, dark haloes, about ten times bigger than the bright galaxies (which, once again, ties in very nicely with other recent studies of galaxies).

Some of the redshifts corresponding to these lines of heavy elements are between about 0.5 and 0.8, much smaller than those of the lines in the Lyman forest, and putting the galaxies responsible for the lines at the edge of the limit of observation with ground-based telescopes. With the evidence of the lines showing them where to look, astronomers have been able to identify very faint, distant galaxies with the right redshifts in many of these cases. Enough of these galaxies have now been identified to establish beyond doubt that the heavy-element lines are indeed produced in galaxies. So, without going through the tedious process of identifying every such galaxy, we can confidently assume that each redshift in this range determined from a heavy-element line in a quasar spectrum tells us the location of a distant galaxy. And, pushing our luck only a little, we can reasonably assume that the heavy-element lines corresponding to still higher redshifts, where we have no hope of identifying the galaxies involved directly, also tell us where galaxies were located when the Universe was young. The obvious question is, how are those galaxies distributed through space?

These studies are still in their infancy, but the most striking feature of this new view of the distant Universe is a huge concentration of galaxies forming a wall 30 million light-years thick and 300 million light-years across at a redshift of about 2, corresponding to a time when the Universe was just 3 billion years old. This super-supercluster, identified only in 1986, is now coming under scrutiny, and studies of features like this at such large redshifts will help to provide a picture of the way the Universe has evolved.

Into the Past

Our overall understanding of galaxy formation and evolution is now at a primitive stage—rather like our understanding of stars fifty years ago. We don't fully understand the most basic structural features—why some galaxies are discs and some are elliptical—though this basic classification was established by Edwin Hubble in the 1940s. The galaxies that Hubble observed were all within a few hundred million light-years of us; relatively close compared to the distance we can now probe. But because the Universe is much the same everywhere, Hubble got a view of a fair sample of it. His classification of galaxies has survived and stood the test of time. Hubble himself, however, was acutely aware of observational limitations, and his great book, *The Realm of the Nebulae*, concludes with these words:

> With increasing distance our knowledge fades, and fades rapidly. Eventually we reach the dim boundary, the utmost limits of our telescope. There we measure shadows, and we search among ghostly errors of measurement for landmarks that are scarcely more substantial. The search will continue. Not until the empirical resources are exhausted need we pass on to the dreamy realm of speculation.

This search *has* continued, as more powerful telescopes and more sensitive detectors have been employed. Observers have invaded the speculators' territory. Because light travels at a finite speed, we see distant parts of the Universe as they were long ago. We can sample the past, even if we cannot repeat it, and we are struck by the realisation that the simple laws of physics derived from experiments here on Earth can be applied across vast reaches of cosmic space and time. To see any cosmic evolutionary trend, however, one must look back in time by a good fraction of the 10 billion-plus years for which the Universe has been expanding. The first person to do this was Sir Martin Ryle, in Cambridge, in the late 1950s. He found clear evidence that conditions were different when galaxies were young. His radio telescopes picked up electromagnetic waves from some active galaxies (the kind that we now think harbour supermassive black holes) even when these were too far away to be observed by the optical techniques of the time. He couldn't determine the distances to such galaxies by radio measurements alone, but he assumed that, at least on average, the ones that seemed fainter were really more distant than those that produced a more intense signal in his instruments. He counted the numbers of radio sources with various apparent intensities and found that there were too many faint galaxies (in other words, more-distant ones) compared to brighter, closer ones. This was discomfiting for those who supported the idea of an unchanging, steady state universe—cosmologists with whom Ryle was, at the time, involved in a running battle. But the observations were compatible with an evolving universe, if galaxies were more prone to violent outbursts in the remote past.

Optical astronomers joined this enterprise after the discovery of quasars in 1963. Because quasars are the hyperluminous nuclei of galaxies, optical astronomers have now seen some so far away that the light set out when the Universe was less than one-fifth of its present age. It is clear from quasars, as it was from Ryle's radio data, that the cosmic scene was much more violent when galaxies were young. Most of the runaway catastrophes, the formation of great black holes, happened early in galactic history, when less gas was locked up in stars and more was still available to fuel the central monsters.*

---

*It is an antianthropic irony that the most interesting time to have been an astronomer was at that epoch, before the Earth formed.

Galaxies at very high redshifts can be studied indirectly by using quasars as probes, as we have explained, and looking for absorption lines, or even gravitational lensing, due to galaxies along the line of sight. These ordinary galaxies, those without hyperluminous quasar nuclei, would be almost invisibly faint at such great distances. But recently the prospects of directly detecting them have brightened; the latest sensitive detectors, such as charge coupled devices (CCDs), reveal huge numbers of objects, closely packed over the sky, which are probably young galaxies at the stage when a cloud of gas is still contracting to form a disc. We must await the launch of the space telescope, and the next generation of ground-based telescopes, to image these objects. The first instrument that we expect to be up to the task will be the ten-metre Keck Telescope in Hawaii; it should produce bright images of these objects that will reveal their shape. We shall then be able to obtain "snapshots" of groups of galaxies at different distances (and therefore different evolutionary stages) and trace directly how galaxies emerged from amorphous beginnings—an initially smooth and almost featureless universal soup.

Almost, but not quite, featureless; there were (although we do not know why) small fluctuations from place to place in the expansion rate. We think we know what happened then. Embryonic galaxies were slightly overdense regions whose expansion lagged behind the average expansion. These embryos eventually evolved into distinct clouds, whose internal expansion halted and went into reverse. The bigger ones collapsed to make the first individual galaxies when the Universe was perhaps 10 percent of its present age. Less-massive systems would have survived as stable gas clouds—these are responsible for the Lyman-alpha forest. Subsequently, the galaxies would have grouped into clusters. This, at least, is the scenario our theories suggest—

only further observations will tell whether theorists' confidence in the story is well based.

The fact that the Universe is simple enough for us to understand it is a Deep Truth. How exactly, though, do we come to be here puzzling over the nature of the Universe? Our existence depends on the production of elements heavier than hydrogen and helium. Lines corresponding to heavy-element absorption are seen at redshifts as high as 3.3 in quasar light, and this alone shows that some stars had already gone through their life cycles and spread their products through young galaxies at that early time. But, as we hinted in the first chapter of this book, those stars were able to process hydrogen and helium into heavier elements only because of an astonishing coincidence involving the energetics of carbon nuclei. Armed now with an understanding of at least the broad picture of the Universe we live in, the time has come to take a closer look at this cosmic coincidence, and others. Is the Universe really tailor-made for humankind?

PART THREE

---★---

The Bespoke
Universe

CHAPTER TEN

★

Tailor-made
for Man?

JUST AS THE ELECTRONS in an atom can occupy different energy levels, like steps on a staircase, so can the particles that make up the nucleus of an atom. The nucleons may change from a low-energy state to a high-energy state, provided they are given the right push (the right amount of energy) from outside; once they are in a high-energy state, they may fall back to a lower level, most probably the bottom step on the energy ladder, and radiate the appropriate amount of energy in the process. For the Universe at large, the greatest cosmic coincidence is that the Universe is so precisely flat, and that the amount of baryonic matter in the Universe is so close, compared with what it might have been, to the amount of dark stuff. Within that flat Universe, however, there is another coincidence almost as remarkable, which allows for the existence of carbon and heavier elements, and which depends on fine-tuning in the energy levels of a handful of atomic nuclei.

One of the reasons why the standard model of the Big Bang is regarded as a scientific triumph is that it explains the abundances of the lightest elements, revealed by spectroscopic studies of gas clouds and of old stars. These are hydrogen, the lightest element of all, helium, which makes up some 25 percent of the baryonic matter, and small traces of deuterium (heavy hydrogen) and lithium. The standard model explains how these elements were made out of primordial baryons during a period from about one-tenth of a second after the moment of creation up to about four minutes later. It doesn't matter how the Universe got to be in the state of heat and density that existed at an "age" of 0.1 second; it would then have a temperature and density high enough to establish a state of "thermal equilibrium" that washed out all trace of its past history, so none of the debate in recent years about the very early Universe and the moment of creation itself affects these calculations. No baryonic material was processed by the Big Bang into elements heavier than lithium, which has three protons and four neutrons locked away in each nucleus. So where did everything else come from?

In fact, this was a well-recognised problem by the 1950s, long before the complete and detailed standard model of the Big Bang had been worked out. The fact that we exist shows that carbon and other elements have been manufactured somewhere and dispersed through space. Even the earliest detailed studies of Big Bang physics, carried out by George Gamow and his colleagues in the United States, showed the difficulty of making anything except hydrogen and helium (a difficulty that the ebullient Gamow used to dismiss by pointing out that since his theory could explain the nature of 99 percent of the known matter in the Universe, the hydrogen and the helium, it must be counted a success!). The only place that heavier elements could have been manufactured, and could still be manufactured today, is inside stars. But how do stars do the trick?

The Beryllium Bottleneck

Astrophysicists knew that the trick, nucleosynthesis, must have something to do with sticking helium nuclei together. The most stable form of helium, helium-4, contains just two protons and two neutrons in its nucleus. This is such a stable configuration that a helium nucleus behaves like a particle in its own right, and was known, as the alpha particle, before neutrons themselves were discovered. Because the helium-4 nucleus is so stable, atoms that are made up, in effect, of whole numbers of helium-4 nuclei are themselves stable, and therefore common, compared with other nuclei. Carbon, which contains twelve nucleons, and oxygen, with sixteen, are the two most obvious examples, with great importance for life forms like us. Once carbon and oxygen exist in the Universe in the right quantities, it is relatively easy, according to the laws of physics derived from studies of, for example, the way alpha particles interact with nuclei in particle accelerators, to build up the heavier elements.

This happens, in essence, by adding alpha particles (helium nuclei) to existing nuclei, which then, sometimes, spit out the odd proton or neutron to produce a nucleus of a slightly lighter element. But there seemed to be a bottleneck at the very first step in this process.

Two alpha particles that collide with each other with the right energy (enough to overcome the electrical repulsion produced by the positively charged protons they each carry) will stick together to form a nucleus of beryllium-8. Unfortunately, however, beryllium-8 is the exception to the rule that nuclei containing whole numbers of alpha particles are stable. It is spectacularly *un*stable, and breaks apart into lighter particles within a lifetime of only 10^{-17} seconds. So how can carbon, which requires the addition of another alpha particle to a beryllium-8 nucleus, ever be built up?

Maybe, some theorists speculated, carbon-12 could be made directly inside stars, when *three* helium-4 nuclei just happened to collide with one another simultaneously. But a simple calculation soon showed that this is indeed about as unlikely a prospect as it sounds. It might happen occasionally, but not often enough to produce all the carbon we see around us, the key element in the chemistry of living things.

In 1952, Ed Salpeter, an American astrophysicist, suggested (more or less in desperation) that carbon-12 might be produced in a very rapid two-step process, with two alpha particles colliding to form a nucleus of beryllium-8, which was then in turn hit by a third alpha particle in the 10^{-17} seconds before it had time to disintegrate. Since this did at least give 10^{-17} seconds for the third particle to arrive, instead of requiring three to meet simultaneously, it was an improvement on the triple-collision idea. But since the arrival of a third particle might very effectively smash the unstable beryllium-8 nucleus to bits, it wasn't much of an improvement. Then, Fred Hoyle, who had, back in 1946, written a classic paper expounding the idea that the chemical elements were made in stars, entered the story.

Hoyle's Anthropic Insight

Hoyle (now Sir Fred) was based in Cambridge, England, but in the 1950s spent time in California, working with his friend, nuclear physicist Willy Fowler. Hoyle puzzled over the problem of how heavy nuclei might be built up in stars (stellar nucleosynthesis), and became intrigued by the possibility that the energy levels of beryllium, helium, and carbon might be just right to encourage the two-step reaction Salpeter had proposed. It all hinged on a property known as resonance.

Resonance works like this. When two nuclei collide

and stick together, the new nucleus that is formed carries the combined mass-energy of the two nuclei, plus the combined energy of their motion, their kinetic energy (and minus a small amount of energy from the strong force, the binding energy that holds the new nucleus together). The new nucleus "wants" to occupy one of the steps on its own energy ladder, and if this combined energy from the incoming particles is not just right then the excess has to be eliminated, in the form of leftover kinetic energy, or as a particle ejected from the nucleus. This reduces the likelihood that the two colliding nuclei will stick together; in many cases, they simply bounce off each other and continue to lead their separate lives. If everything meshes perfectly, however, the new nucleus will be created with exactly the energy that corresponds to one of its natural levels (it can then, of course, emit packets of energy and hop down the steps to the lowest level). In that case, the interaction will proceed very effectively, and the conversion of lighter nuclei into a heavier form will be complete. This matching of energies to one of the levels appropriate for the new nucleus is the effect known as resonance, and it depends crucially on the structure of the nuclei involved in the collisions.

In 1954, Hoyle realised that the only way to make enough carbon inside stars is if there is a resonance involving helium-4, beryllium-8, and carbon-12. The mass-energy of each nucleus is fixed and cannot change; the kinetic energy that each nucleus has depends on the temperature inside a star, which Hoyle could calculate. Using that temperature calculation, Hoyle predicted that there must be a previously undetected energy level in the carbon-12 nucleus, at an energy that would resonate with the combined energies, including kinetic energy, of its constituent parts, under the conditions prevailing inside stars. He made a precise calculation of what that energy level must be, and he cajoled Willy

Fowler's somewhat sceptical nuclear physics colleagues until they carried out experiments to test his prediction. To the astonishment of everyone except Hoyle, the measurements showed that carbon-12 has an energy level just 4 percent above the calculated energy. This is so close that the kinetic energies of the colliding nuclei can readily supply the excess. This resonance greatly increases the chances of a helium-4 and a beryllium-8 nucleus sticking together, and ensures that enough alpha particles can be fused into carbon nuclei inside stars to account for our existence.

The remarkable nature of Hoyle's successful prediction cannot be overemphasised. Suppose, for example, that the energy level in carbon had turned out to be just 4 percent lower than the combined energy of helium-4 and beryllium-8. There is no way that kinetic energy could *subtract* rather than add the difference, so the trick simply would not have worked. This is made clear when we look at the next putative step in stellar nucleosynthesis, the production of oxygen-16 from a combination of carbon-12 and helium-4. When a carbon-12 nucleus and a helium-4 nucleus meet, they would fuse into oxygen if there were an appropriate resonance. But the nearest oxygen-16 resonance has one percent *less* energy than helium-4 plus carbon-12. But that 1 percent is all it takes to ensure that this time resonance does not occur. Sure, oxygen-16 is manufactured in stars, but only in small quantities (at least, at this early stage of a star's life) compared with carbon. If that oxygen energy level were 1 percent lower, then virtually all the carbon made inside stars would be processed into oxygen, and then (much of it) into heavier elements still. Carbon-based life forms like ourselves would not exist.

Most anthropic arguments are made with the benefit of hindsight. We look at the Universe, notice that it is close to flat, and say, "Oh yes, of course, it must be that way, or we wouldn't be here to notice it." But Hoyle's prediction is different, in a class of its own. It is a genuine scientific prediction, tested and confirmed by *subsequent* experiments. Hoyle said, in effect, "since we exist, then carbon must have an energy level at 7.6 MeV." *Then* the experiments were carried out and the energy level was measured. As far as we know, this is the only genuine anthropic principle prediction; all the rest are "predictions" that *might* have been made in advance of the observations, if anyone had had the genius to make them, but that were never in fact made in that way.

Hoyle's remarkable insight led directly to a detailed understanding of the way in which all of the other elements are built up from hydrogen and helium inside stars. He worked closely with Willy Fowler on this, and with the husband-and-wife team Geoffrey and Margaret Burbidge. Fowler (without Hoyle) later received a Nobel Prize for his part in the study of stellar nucleosynthesis.

This combination of coincidences, just right for resonance in carbon-12, just wrong in oxygen-16, is indeed remarkable. There is no better evidence to support the argument that the Universe has been designed for our benefit—tailor-made for man. But there are alternative ways of viewing this coincidence, and others. So before we present the alternative view we should perhaps mention at least two other striking coincidences that help to make the Universe a fit place for life.

The Stellar Pressure Cooker

Making carbon, and heavier elements, inside stars solves only half the problem of how carbon-based life forms come to be here on Earth, puzzling over their origins. How do the heavy elements get out of the stars and spread across the Galaxy to become part of the clouds of material from which new stars and planets form? The simple answer is that the heavy elements are spread when a minority of stars explode as supernovae. But what makes a supernova blow its top? It turns out that this spreading of the stuff of life across the cosmos also hinges upon a close cosmic coincidence.

Because of the failed resonance at oxygen-16, life for a very massive star is both complicated and, ultimately, disastrous. All stars start their lives by "burning" hydrogen nuclei, converting them into helium and releasing heat in the process. When hydrogen is exhausted, helium in its turn can be burned to produce carbon—at this stage of its life, a star like our Sun swells up to become a red giant. As long as helium is being converted into carbon, with a net release of energy for each carbon nucleus formed, the star can stay hot enough in the centre to support the weight of its outer layers. But eventually, after many millions of years, the helium is exhausted. What happens next depends on the mass of the star. By far the majority of stars run through some further nuclear reactions, in a last-ditch attempt to maintain their former glory, then collapse and cool down, huddling in upon themselves to form a ball of dead star material, a white-dwarf star that may have the mass of our Sun but that occupies a volume no bigger than the Earth.

In the later stages of its life, such a star may puff away large amounts of material into space. But this material comes only from the outer layers of the star and is almost entirely hydrogen and helium. These stars are of no interest to our story, because the heavy elements they have manufactured stay locked up in the dead star. But the first stages in the story of a supernova follow the same path as these more mundane objects, which goes something like this. When helium burning ends, the compression of the inner core by the weight of the outer layers of the star forces temperatures up beyond 600 million K, to the point where carbon burning begins. These central temperatures are reached only in stars more than four times as massive as our Sun; lesser stars almost immediately, once their helium is burnt, settle down quietly into a respectable old age. Because of the lack of a suitable oxygen resonance, carbon burning in more massive stars is not simply a matter of adding an alpha particle to a carbon-12 nucleus to make oxygen-16. Carbon burning actually involves collisions between pairs of carbon-12 nuclei. Two carbon-12 nuclei may collide, with one alpha particle being ejected (carrying off excess kinetic energy) to produce a nucleus of neon-20. Sometimes, the two carbon nuclei stay stuck together, as magnesium-24, with excess energy removed by a burst of gamma rays; and sometimes *two* helium-4 nuclei are ejected from the merging pair of carbon-12 nuclei, leaving oxygen-16 behind.

This set of processes can continue only as long as there is carbon available. When it is exhausted, the gravitational collapse of the star sets in again. Once more, many stars end their lives at this stage and settle down into cooling lumps of matter. But if the mass of the star exceeds nine solar masses, the temperature rises above a billion degrees, and neon burning begins. This process rearranges the leftover material from carbon burning. One neon-20 nucleus may absorb a helium-4 nucleus to become magnesium-24, while ejecting a gamma ray; another neon-20 nucleus will absorb a gamma ray and spit out helium-4, leaving oxygen behind.

As the temperature rises above 1.5 billion K, oxygen burning begins. This is an even more complicated process, since when two oxygen-16 nuclei collide they can produce a variety of elements, including two isotopes of silicon, two forms of sulphur, phosphorus, and more magnesium. The key ingredient here is silicon, because at the next stage of nuclear burning, at a central temperature of 3 billion degrees kelvin in a star with a mass twenty or more times that of our Sun, silicon becomes involved in a series of hundreds of nuclear reactions, which yield as their end product the ultimate stellar ash, in the form of nuclei of iron-56.

It sounds a long and complicated story, and in a sense it is. But by the standards of stellar lifetimes, the later stages are all over in the blink of an eye. Even one of these massive stars may spend tens of millions of years burning hydrogen quietly, and shining like our Sun, then millions of years more as a helium-burning red giant. But in a star with, for example, twenty-five times the mass of our Sun, carbon burning takes just six hundred years, neon burning lasts for about a year, and oxygen burning is over in six months.* The final stage, silicon burning, is over in less than a day.

There are three reasons for this frantic speeding up of stellar processes. First, at each stage the star collapses inward a little more; its core gets hotter, and the nuclear reactions become more violent. Second, the later nuclear fuels are less efficient. A single nucleus of silicon 28, for example, weighs almost as much as 28 of its hydrogen nuclei (protons) the star began with, but can release far less energy. Third, such high temperatures are reached that neutrinos can be created; these escape freely, removing energy far faster than the stellar surface radiates.

Energy is the key to what happens next. All the energy the star has derived from nuclear reactions comes from packing protons and neutrons more tightly to-

*All the figures in this section are based on computer models, which calculate how stars evolve subject to the known laws of physics. They are calibrated by comparing the computer "predictions" with the appearances of different kinds of real stars, with different ages and masses. The figures are consequently quite reliable.

gether in atomic nuclei. In iron-56, they are packed as tightly together as possible, and no more energy can be provided by fusion. More massive nuclei, such as gold, lead, silver, and uranium, are less tightly packed than iron-56. To make them out of iron, *more* energy has to be put into the system. This, among other things, is what happens in a supernova.

The Supernova Connection

For the sake of argument, we will carry on describing what happens to a star with a mass of about twenty-five suns after silicon burning is complete and it is left with a ball of iron, about as massive as our Sun, in its centre. Only the details are different for stars with different masses. The star now has no effective means of support, since there is no more energy being produced by nuclear burning in its core. The result is dramatic. The inner regions of the star are squeezed inwards, and the pressure on the iron nuclei in the core becomes so great that electrons and protons are forced to merge into one another, forming neutrons. A ball of neutrons can indeed pack matter together more compactly than a ball of iron nuclei, and the centre of the star begins to convert into a neutron star, still with as much mass as our Sun but now occupying only as much space as Mount Everest. It becomes, in effect, a single "atomic" nucleus. Material from the inner part of the surrounding star has the floor pulled from underneath it, and plummets down onto the newly forming neutron star, reaching speeds as great as 15 percent of the speed of light. When this fast-moving material hits the neutron star, from all sides at once, the shock actually squeezes the ball of neutron material, like a golf ball being squeezed in an iron grip. But neutron stuff is very difficult to compress—it is like trying to squeeze the nucleus of an atom—and quickly bounces back from this compression. Enormous pressures and temperatures are created in the bounce, which turns the shock wave inside out and sends it speeding back out through the giant star.

Everything has happened in less than half a second. As the shock begins to move outwards through the star, which may have a diameter of 700 million kilometres, as big as the orbit of Jupiter, it encounters resistance and begins to slow down. It is, after all, trying to move bodily about twenty-four solar masses of material. Without help, it would fizzle out. But it is followed by a flood of neutrinos produced in the neutron core of the star when it was squeezed by the infalling matter. The matter in the slowing shock wave is so dense that it actually absorbs a significant number of neutrinos. The energy from the neutrinos gives the shock wave the boost it needs to finish the job of blowing apart the outer layers of the star.

In all this energetic activity, quantities of elements heavier than iron have been formed, and many complex nuclear interactions have produced a variety of other elements from the basic products of nuclear burning. A supernova shines, for a few weeks, as brightly as a whole galaxy of normal stars put together, and the energy that makes it so bright comes from radioactivity, from unstable elements heavier than iron that were put together by the shock wave and are now breaking apart, releasing energy and forming more stable, tightly packed nuclei. From the site of this intergalactic beacon, much more than twenty times the mass of our Sun, in our specific example, is expelled completely into space, and carries with it this heavy-element legacy from the dying star. The core, at last free from the burdensome pressure of the rest of the star, settles down as a spinning neutron star, perhaps to be detected by some civilisation as a pulsar. The organic beings that study such pulsars, and the steel girders of which their radio telescopes are manufactured (not to mention the silicon in the chips of their computers), are equally the products of supernova explosions in aeons gone by.

The story is fascinating in its own right. But where is the anthropic coincidence? It lies in that burst of neutrinos, the crucial step in helping the shock wave to blow the star apart. Computer calculations in the 1980s had shown that the shock wave alone simply could not do the job, and that neutrinos must be involved. But some researchers were sceptical, because the properties of neutrinos had to be precisely "fine-tuned" to do the job. It all hinges upon the strength of the weak interaction, one of the four fundamental forces of nature. This is the force that determines how strongly neutrinos interact with baryons. If the weak interaction were a little too weak, then even the dense shock wave would be transparent to neutrinos, and they would flood out through the star without getting involved in pushing apart the outer layers of the star. If, on the other hand, the weak interaction were a little too strong, then the neutrinos would get involved in reactions in the core itself, and would never get out to the region where the shock wave was slowing down and giving up the ghost. The weak interaction has to be just right to allow enough neutrinos both to escape from the core and to interact with the shock wave.

Some doubts about this scenario for the explosion mechanism were allayed by studies of the burst of neutrinos from supernova 1987A. The energy of these neutrinos, and the implication, from the arrival of a handful in our detectors, of how many escaped from the core of the supernova, match the requirements of the models. Studies of the supernova match very well with the computer calculations, supporting the view that neutrinos are indeed the driving force in expelling large quantities of gas enriched with heavy elements into space—a phenomenon without which no planets like Earth or creatures like us would exist.

A Cosmic Connection

The same coincidence crops up earlier in the life of the Universe. It is the strength of the weak force that decides how much hydrogen is processed into helium in the Big Bang. It requires a rather precise fine-tuning to avoid a runaway in one direction or the other—make the force slightly stronger and no helium would have been produced; make it slightly weaker and nearly *all* the baryons would have been converted into helium in the Big Bang. A universe in which stars were initially made only of hydrogen might not be so very different from our own; but if all the stars were originally composed of helium, they would have burnt out more quickly, perhaps not giving life time to evolve on any planets that formed (if life can indeed develop without hydrogen present to form water). The condition that some stars go through a supernova phase (triggered by a neutrino-boosted shock) is essentially the same as the condition that there be an interesting amount of cosmological helium production. The weak force seems to be just about as weak as it can be in order to avoid all the original hydrogen being converted into helium. Supernovae might still work (exploding by a different mechanism) if the force were a little stronger, but if the force were weaker the neutrinos could not drive any kind of explosion; the Universe would be even more comfortably dominated (baryonically speaking) by hydrogen if the force were a little stronger. But the window of opportunity for a universe in which there is *some* helium, *and* exploding supernovae, is very narrow.

These examples are enough to demonstrate the power of the coincidences at work in our Universe. But there is another level at which we can contemplate the puzzle of our existence. So far, we have taken the framework of our Universe for granted. We have talked of tinkering with the weak force, or the constant of gravity, and we have happily discussed bending and stretching spacetime. But is the fabric of the Universe itself something unique and special? Can we read any significance into the fact that we live in a world built up from three dimensions of space and one of time?

Space, Time, and the Universe

Take time first. At the Planck time, we have to jettison the whole idea of an arrow of time, and even that there are three dimensions of space and one of time. We can ask *why* there was a Big Bang—is there something uniquely self-consistent about the way it happened, or did it have coincidental features without which we would not be here—but we cannot ask what happened "before" the Bang. There's a sense in which time itself *begins* with the Big Bang. Similarly, time will end if there is a big crunch; and even if the Universe expands forever, time would end for any observer who is swallowed by a black hole and pulled into the central singularity.

All this assumes that time is measured by the ticking of standardised clocks. But we face a conundrum if we try to imagine how we could measure the time that elapsed before there was a big crunch. Any conceivable clock would be destroyed at sufficiently high density. We might imagine starting off measuring time in years, by orbits of planets around stars. When the density gets high and solar systems are destroyed, we could use an atomic clock. But eventually atoms themselves would be destroyed. In approaching the singularity we would need to rely on an infinite sequence of successively smaller and sturdier clocks.* This line of thinking, however, is based on an infinite regress that is almost certainly unwarranted—time is *not* infinitely divisible.

Just as there may be a limit, measured in tens of billions of years, to the longest timespan that is meaningful (the time from Big Bang to big crunch), so there may be a smallest natural unit of time. Conventional physics does set such a limit. The Heisenberg uncertainty principle tells us that if we want to measure a short time interval with increasing precision, we need to use quanta of radiation with more energy, and shorter wavelengths. Because the light quanta move at finite speed, this increasing amount of energy must be focussed into a smaller and smaller space. A limit arises when the concentration of energy is so great that the quantum collapses into a black hole. This occurs at about 10^{-43} seconds, the Planck unit of time. It is not possible, according to quantum theory, to place events in chronological order with any greater precision than this; some physicists suggest that there may actually be a fundamental limit bigger than the Planck time, although experiments tell us that the graininess of time is certainly not on scales larger than 10^{-26} seconds.

Another fundamental mystery is "time's arrow," the irreversibility of the flow of time, from the Big Bang into the future. The laws of microphysics are reversible in time, and if we took a movie of microphysical interactions and ran it backwards we would not, in general, be able to tell that time had been "reversed." The

---

*This reminds us of Zeno's paradox, "proving" that motion is impossible: before an arrow can reach its target, it must first get halfway there; before that, it must get a quarter of the way; before that . . . and so on, in an infinite number of steps that must be completed before the arrow can wing its way.

macroscopic world is by no means time-reversible in this sense. Things age and wear out—and this ageing and wearing out of things is described by physical laws of thermodynamics. We have memories of the past only, not the future, and in general it is easier to be wise after the event than to predict the future.

The same arrow of time also seems to apply to the expanding Universe, with the Big Bang in the past. "Later" times are times when clusters of galaxies have moved farther apart, and this is a universal arrow of time that could be deduced by intelligent observers anywhere in the Universe. What would happen if the Universe were to halt its expansion and begin to recollapse? The big crunch, the counterpart to the Big Bang, would then lie in the future, not the past. "Later" times would be times when clusters of galaxies were closer together—or would they be? Thomas Gold, in the 1960s, was one of several theorists who speculated that if this happened time would run backwards. Does that mean we would "remember" the future? Would intelligent observers in a collapsing universe still think that "later" times were when clusters were farther apart?

This conjecture seems highly implausible. Nothing special happens locally, anywhere in the Universe, at the epoch of maximum expansion. Nevertheless, the idea has been taken seriously, and has been revived in the 1980s by Stephen Hawking and other researchers trying to develop a quantum-mechanical description of the Universe. Paul Davies, of the University of Newcastle upon Tyne, has recently suggested that this casts doubt on one of the most fundamental laws of physics.

This is the second law of thermodynamics, which states that disorder (entropy) is always increasing. Hawking showed that a black hole possesses entropy, and that the area of the surface of the hole is a measure of its entropy. If entropy can only increase, that means that a black hole can only get bigger (except when it is evaporating, a quantum process that itself increases entropy). If the Universe is closed, to form the three-dimensional equivalent of the surface of a sphere, then in many ways it resembles a black hole viewed from the inside. But if the Universe is destined one day to have its expansion stop and recollapse, that implies that the "area" of the "black hole" will one day shrink—that entropy decreases and the second law of thermodynamics does indeed break down, in regions of space and time that are as normal as the Universe today. But there may be a way out of this dilemma. Roger Penrose, of the University of Oxford, has developed the idea that time's arrow is related to the difference between the dynamics of the Universe in the Big Bang and the big crunch. For reasons that are not yet understood, the Universe emerged from the Big Bang in an amazingly smooth state. Penrose argues that this makes the initial singularity a special and unusual one. The big crunch, on the other hand (if it happens), will be much more messy, chaotic, and unsynchronised. Crumpled regions of spacetime that have already formed black holes in the Universe will clump together, crumpling space still more as the Universe collapses. Penrose believes that there may be a law of physics—which nobody has yet formulated—according to which past singularities are always simpler in structure than future singularities, and that this is why we perceive time as flowing from the Big Bang into the future of the expanding Universe.

Another enigma concerns closed loops in time. If it were possible to traverse such a loop and come back in your own past, there would be immediate and obvious scope for contradictions. Strangling your own grandmother in her cradle poses problems not merely of ethics but of causality—where did the strangler come from, if granny never grew up? Astoundingly, there are some solutions to Einstein's equations of general relativity that appear to represent cosmological models that permit loops in time. These possible universes (explored mathematically by Kurt Gödel in the 1940s) differ from the structure of our own Universe, but they do not seem physically impossible.

One can take either of two attitudes to these closed loops. On one hand, the fact that Einstein's equations permit them may be telling us that general relativity is incomplete, and an extra law of nature is needed to rule out such absurdities. On the other hand, one could take the view that although there are clear paradoxes for conscious observers who travel around such loops in time, they entail no obvious absurdity in a universe where there is no memory. A Gödel universe might, by definition, be devoid of intelligent life—a kind of antianthropic requirement—although, in any case, since it would take almost infinite energy and a time nearly as long as the age of that universe to travel around one of these loops, they could do no practical harm. Nevertheless, physicists react against any model that permits causality to be violated. Most physicists feel equally strongly about theories that envisage the arrow of time going into reverse—but, again, this might not entail any logical absurdity, if it happened at a time when all stars and matter had decayed, black holes had evaporated, and the Universe contained nothing except pure radiation, with no conscious observers around to notice what was going on.

There is plenty of scope, it seems, for speculation about the nature of time. But there is far less scope to speculate on the possibility of universes constructed out of a different number of dimensions from our own. The fact that our Universe is composed of (three plus one) dimensions is actually such a profound observation that it caused at least a few people to begin thinking along anthropic lines long before the idea of the anthropic cosmological principle itself was articulated. Although he did not express it quite in those terms, one of the first people to follow through the implications of the three-dimensionality of space was William Paley, an eighteenth-century philosopher and churchman.

Paley is best remembered today for his forceful expression of the argument that living things are far too complicated to have arisen by chance, and that the existence of creatures as beautifully fitted for their way of life as ourselves (or a fly, or a primrose) reveals the presence of a designer, the hand of God at work. This argument is expressed dramatically in terms of a man who finds a watch lying upon the ground, and who knows nothing about watches but perceives from an inspection of the object that it has been designed and put together for a purpose. A "blind watchmaker" who sat before a heap of watch components and put them together *at random* could never, the argument runs, assemble a working watch.* The argument may be correct, but it is inappropriate, since evolution by natural selection does not proceed by sticking together all the components of a living creature at random, but by building step by step on previous successes. That debate is outside the scope of our present book, but Richard Dawkins has laid the myth to rest in his superb *The Blind Watchmaker*, which we recommend to anyone still seduced (or confused) by the "argument from design." What *is* relevant to our main theme is that Paley was also intrigued by the inverse square law of gravity, described by Newton in the 1680s.

Paley realised that the inverse square law, in which the force between two objects is proportional to 1 over the square of the distance between them, is unique

---

*Paley's arguments came mainly from biology, but as astronomers we are amused to note the rather limited relevance he assigned to our subject: "My opinion of astronomy has always been that it is not the best medium through which to prove the agency of an intelligent creator, but that this being proved, it shows beyond all other sciences the magnificence of his operations."

in giving rise to stable orbits. If the law of gravity had, for example, been an inverse *cube*, then planetary orbits would be unstable, and a planet that moved a little closer to the Sun would immediately begin to fall inwards permanently, while one that moved slightly outwards in its orbit would continue receding forevermore. Tiny changes, such as those caused by the impact of a meteorite, would be disastrous. In our Universe, if the Earth's orbit, say, shifts slightly inwards or outwards because it is hit by a piece of rock from space, the natural tendency is for the planet to return close to its old, regular path. Paley saw this "choice" of the inverse square law of gravity as another example of God's work in designing a Universe suitable for human life. He did not elaborate, however, on the fact that the inverse square law is a byproduct of the fact that the Universe has three spatial dimensions—although this had been noticed by Immanuel Kant earlier in the eighteenth century.

The importance of the dimensionality of space began to interest scientists in the twentieth century, following Einstein's work that gave space (or spacetime) a dynamic role in physics. This showed that the dimensionality of the law of gravity is always one less than the dimensionality of space—inverse square in a space of three dimensions, inverse cube if space has four dimensions, and so on. Planetary orbits are stable only in a space with three dimensions, because an inverse square law of gravity is the natural law only in a space with three dimensions. About the same time, researchers realised that the equations of electromagnetism, discovered by the Scotsman James Clerk Maxwell in the nineteenth century, have workable solutions only if they are applied in a spacetime that has (three plus one) dimensions.

The insight was developed by G. J. Whitrow in 1955. He argued that the reason we observe the Universe to possess three dimensions is that observers can exist *only* in universes that have three dimensions of space (and one of time). Life can exist only in three-dimensional space; we live, and therefore it is no surprise that we find ourselves in three-dimensional space. But in the 1950s, the way people thought about the implications of this insight was beginning to be quite different from the way Kant or Paley thought about the significance of the inverse square law. Instead of regarding our Universe as unique, and seeing its suitability for life as the result of design, some researchers began to consider the possibility that there might be many universes, in an array (an "ensemble") that included all possibilities of, for example, dimensionality. All possible worlds exist, in this picture, but life exists only in the subset of worlds suitable for life, and there is no need to invoke a designer at all. Hardly surprisingly, some of the first discussions of this idea came from researchers in the Soviet Union, where the hand of God was not regarded as a reasonable explanation for the cosmic coincidences; Fred Hoyle also considered the possibility that the energy levels in carbon, for example, might follow different laws in different parts of the Universe, or the superuniverse, so that life like us exists only in lesser regions where the coincidences work out just right (but, unlike the Soviet researchers, Hoyle has developed his own ideas about the hand of a designer at work in setting those coincidences).

Some people now take this very seriously. John Wheeler, for example, imagines an ensemble of universes with different physical laws and different values of the fundamental constants, all "laid down" at the initial singularity, the moment of creation. A kind of evolution by natural selection operates on these universes. Most of them are "stillborn," in the sense that the prevailing laws do not permit anything interesting to happen in them. But maybe in some of them complex structures can evolve. It would be a major achievement if someone could show that any universe in which interesting things can happen *must* end up looking something like our own Universe. At present, though, the best we can do is to show how even a small change in just one of the critical numbers can make a universe that would be almost unrecognisable and probably uninhabitable. There are hundreds of ways in which we could tinker with the laws of nature; but since we have been discussing gravity, let's consider, as our single case study, an adjustment in the strength of that force.

An Alternative Universe

There is nothing truly fundamental about the units by which we choose to measure things here on Earth—pounds or kilograms, yards or kilometres, all are equally arbitrary. So when the more philosophically inclined physicists speculate about the nature of the Universe, they sometimes like to use what they call "natural" units," which are defined in terms of truly fundamental constants of nature, such as the speed of light and the quantum-mechanical constant named after Max Planck (accepting, for the moment, that these constants really are constant). Using these fundamental constants, the strengths of the various interactions can be described in terms of pure numbers, the size of which indicates their importance on a chosen scale. Usually, physicists choose to define these numbers in terms of the mass and electrical charge on a proton, one of the biggest fundamental particles. On that scale, gravity is relatively insignificant, and its strength is described by a number known as the gravitational fine structure constant. This is about 10^{-40}—the electric fine structure constant, in the same system of units, is 1/137, a little less than 10^{-2} and about 10^{-38} times stronger than gravity.

The strength of gravity is tiny compared with the strength of electrical forces, but as we have seen before, it is dominant in the Universe at large, because all the electrical forces cancel out, while all the gravitational forces, from every proton and every other subatomic particle, add together. What would happen if gravity were a bit more dominant?

Consider a universe in which the gravitational fine structure constant is 10^{-30} rather than 10^{-40}, but everything else is as usual. Galaxies, stars, planets, mountains, and microorganisms can all still exist, but they will be very different from their counterparts in our Universe.

The mass of a star depends on the inverse (3/2) power of the gravitational constant (that is, 1 over the square root of the cube of the constant). Lifetimes of stars, on the other hand, depend on the inverse power (1 over) the constant. In our Universe, the Sun is a typical star. It has one solar mass of matter (of course) and a lifetime of about 10^{10} years. So in our alternative universe stars will typically have masses of around 10^{-15} solar masses (about 10^{12} tonnes, roughly the mass of an asteroid in our Universe). The lifetime of a star depends on how long it takes a photon to diffuse (or "random walk") from the centre to the surface. This time depends on the *square* of the distance to be travelled in a straight line. So these ministars, with roughly the same density as stars in our Universe but sizes a hundred thousand (10^5) times smaller, have lifetimes shorter by a factor of 10 billion (10^{10})—about one of our years. From the point of view of the origin of life, involving heavy elements produced in the first generation of stars, the universe will be at an interesting stage when it is about one year old and its Hubble radius is about one light-year; the critical density to make this compact universe flat will be 10^{10} times the density of our Universe—still considerably thinner than thin air, but more substantial than the gas between the stars of the Milky Way.

In the alternative universe, there will be as many galaxies as we see in the observable Universe (about 10^{10}, 10 billion), but each one will be smaller in diameter by a factor of 10^{10} than our Milky Way, and denser by the same factor. Each galaxy will contain about a hundred thousand stars—but the stars will be very different from the ones we know.

With the mass of each star, and hence its energy supply, reduced to 10^{-15} that of the Sun, and a lifetime of one year instead of 10^{10} years, each star will be about 10^{-5} times the brightness of the Sun. These fast-burning stars will be a little hotter than typical stars in our actual Universe, with central temperatures of about 50 million K, compared with the 15 million K at the heart of the Sun. The whole star is about two kilometres across, a tiny nuclear furnace burning hotly with a more blue light, richer in ultraviolet, than the Sun. The energy is available for life to evolve, if these stars have any planets the right distance away.

The right distance is $10^{-2.5}$ times the distance the Earth is from the Sun, because the brightness of the star is down by 10^{-5}, and the intensity of light at any distance from the star depends on the square of the distance, making up the extra factor of 2 in the power of 10. A planet at this distance from its parent star, about 500,000 kilometres (less than twice the distance from the Earth to the Moon) would have a comfortably Earth-like mean temperature of close to 25 degrees C. Just as the parent star has about 10^{-5} times the mass of our Sun, so this "other earth" would have a mass of about 10^{-5} times (one hundred-thousandth of) the mass of our planet. This planet—about the size of a small moon in our Universe—will orbit around its parent star about once every twenty days, by our calendars; this has the curious consequence that in terms of its own "year," that planet "lives" in a universe just eighteen "years" old.*

But a variation on the anthropic arguments that we used in chapter 1 to explain the sizes of planets in our Solar System suggests that there would be a rather large number of "days" in each of those years. If the planet rotates nearly as fast as it can without breaking apart, the number of "days" in each of its relatively long "years" will be something like 2 million. Each "day" on our miniworld is just under one of our seconds long.

Science fiction writers, no doubt, could have a field day describing the culture of a civilisation that inhabited such a planet, and evolved on a timescale set by the length of the day, while the planet orbited the star on a timescale 2 million times longer. Unfortunately, their speculations might have to be described more as fantasy than as genuine scientific fiction, for it is hard to see how life and civilisation could evolve on such planets.

The first problem is that in our hypothetical, compact, speeded-up universe stars will be packed much more closely together, even in comparison with their own size, than stars in our Milky Way. In that universe, the separation between stars is only about 10^{-2} times the distance from the Earth to the Sun, smack in the zone of comfortably warm planetary orbits. Planets in those orbits will be tugged free from their parent star by the gravity of passing stars, and only planets in very close, uncomfortably hot orbits will stay tied to their own star. For those who like such speculation, however, we can envisage habitable planets as tramps be-

*The change in the strength of gravity ought, at first sight, to tie the planet more tightly to its star, and make it orbit faster; but the smaller mass of the star more than compensates for this, leaving a factor of a hundred thousand (10^{-5}) the other way.

tween the stars, wandering here and there but always about the right distance from one star or another to allow the conditions required for complex chemistry to be maintained on their surfaces. What would those surfaces, and any life forms that inhabited them, be like?

The biggest possible mountains would be those just less than the mass required to melt the material at their bases by pressure. They would be just thirty centimetres high. But the mass of a living creature that could just survive falling over without breaking (roughly equivalent to our own mass in the context of the Earth's gravity) would be so small that each living organism could contain no more than about 10^{20} atoms of heavy elements such as carbon, nitrogen, and oxygen. Our own bodies contain about 10^{28} such atoms, ten thousand million times more; the mass of a prospective mountain climber in our alternative universe would be roughly a thousandth of a gram, not a hundred kilograms, and it is very hard to see how such a tiny organism could contain the complexity of chemical compounds that seem to be a prerequisite for intelligent life.*

This is especially unfortunate since if the microorganism could take an intelligent interest in its surroundings it would be able to do astronomy and cosmology in "real time," watching stars and the universe evolve, and living and working for an appreciable fraction of the life of the speeded-up universe.

The relationship between gravity and life revolves

---

*Very hard, but perhaps not quite impossible. A bacterium here on Earth, with a diameter of about a hundred millionths of a metre (one hundred microns), weighs in at roughly half a millionth of a gram. Our hypothetical inhabitants of the miniuniverse are at least a thousand times bigger than that. Science fiction writer Greg Bear explored the possibility of intelligence on the scale of cells like bacteria in his intriguing book *Blood Music*.

around two features of this peculiar force that holds together individual stars and entire galaxies. These features are quite crucial for cosmogonic processes. The first point is that gravity drives things *farther from equilibrium*, not towards equilibrium. When gravitating systems *lose* energy they get *hotter*; for example, an artificial satellite *speeds up* as it spirals downwards due to atmospheric drag. Another example is the Sun. If the heat it loses were not balanced by the release of energy by nuclear fusion in its interior, the Sun would contract and deflate—but it would thereby end up *hotter* inside than before. It needs *more* pressure inside it to balance the stronger tug of gravity when it is more compressed. This runs counter to the general rule of thermodynamics, that hot objects left to their own devices (like a glowing lump of hot steel) radiate heat and get cooler. From the initial Big Bang to our present Solar System, this antithermodynamic behaviour of gravity has been amplifying density contrast and creating temperature gradients—prerequisites for the emergence of any complexity in the Universe.

The second key feature of gravity, in our Universe, is its *weakness*. Our Universe is large and diffuse and evolves slowly *because* gravity is so weak. The extravagant scale of the Universe, billions of light-years, is necessary to provide enough time for the cooking of elements inside stars and for interesting complexity to evolve around even just one star in just one galaxy. There would be less time and less scope for such evolution in the small-scale speeded-up universe discussed above, where gravity is stronger than in ours. A force like gravity is essential if structures are to emerge from amorphous starting points; but, paradoxically, the weaker that force is, the greater and more complex are its consequences.

Although gravity does play this unique role, the exact values of the strengths of the other fundamental forces seem to be just as important for life. The example we have elaborated in detail is typical of such exercises. If we modify the value of one of the fundamental constants, something invariably goes wrong, leading to a universe that is inhospitable to life as we know it. When we adjust a second constant in an attempt to fix the problem(s), the result, generally, is to create three new problems for every one that we "solve." The conditions in our Universe really do seem to be uniquely suitable for life forms like ourselves, and perhaps even for any form of organic complexity. But the question remains—*is* the Universe tailor-made for man? Or is it, to extend that analogy, more a case that there is a whole variety of universes to "choose" from, and that by our existence we have selected, off the peg as it were, the one that happens to fit? If so, what are the other universes, and where are they hiding?

CHAPTER ELEVEN

★

Or Off the Peg?

FINE-TUNING ARGUMENTS have a long history. Lawrence Henderson, a Harvard professor, wrote an important book in this vein, titled *The Fitness of the Environment*, early in the twentieth century. He noted such things as the anomalous property of water, that it does not reach maximum density at the freezing point, which has played a key role in the evolution of life on Earth; he also pointed out special features of other molecules, including carbon dioxide. He argued that "we are obliged to regard this collection of properties as in some intelligible sense a preparation for the process of planetary evolution. Therefore the properties of the elements must for the present be regarded as possessing a teleological character." His contemporary Homer Smith, on the other hand, was unimpressed, saying that "the fitness of the living organism to its environment or vice versa is as the fit between a die and its mould, between the whirlpool and the riverbed."

What, though, can we make of the coincidences in the physical constants involved in nucleosynthesis? They cannot be dismissed as readily as other arguments. A complicated biological organism must indeed evolve in tune with its environment; but the basic physical laws are "given," and nothing can react back to modify them. It does seem worthy of note that these laws permit something interesting to have happened in the Universe, where there could so easily have been a "stillborn" universe in which no complexity could evolve. The Canadian philosopher John Leslie has offered a neat analogy. Suppose you are facing execution by a fifty-man firing squad. The bullets are fired, and you find that all have missed their target. Had they not done so, you would not survive to ponder the matter. But, realising you are alive, you would legitimately be perplexed and wonder why. In its mildest form, anthropic reasoning is simply a proper allowance for observational selection. Given the brute fact that we are a carbon-based form of life slowly evolved around a G-type star, there are some features of the Universe, some constraints on physical constants, which can be inferred quite straightforwardly.

Can we, though, go beyond a subjective expression of surprise that delicate balance seems to prevail? Some distinguished scientists have given their reactions in print, in popular articles if not in technical papers. Freeman Dyson says that in some sense "the universe knew we were coming." And, to quote from Sir Fred Hoyle's *Galaxies, Nuclei, and Quasars*, "the laws of physics have been deliberately designed with regard to the consequences they produce inside stars. We exist only in portions of the universe where the energy levels in carbon and oxygen nuclei happen to be correctly placed."

In response to Hoyle's comment, we might ask if the microphysical constants could be different in different parts of the Universe, or at different times. Paul Dirac suggested, half a century ago, that the gravitational constant might change as the Universe aged; this is now ruled out by observations, and there is no evidence that any other microphysical constants have varied—strict constraints are imposed by observations of the spectra of distant objects, from radioactive decay over the geological past, and other studies; moreover, there are conceptual problems in defining what we would mean by, say, a variation in Planck's constant. The atoms and nuclei whose properties are studied in the laboratory certainly seem to behave identically to those in the most remote quasar or in the first few minutes of the Big Bang. Were this not so—were there no firm link between the cosmos at large and local physics—scientific cosmology could have made little progress (and this book would never have been written).

But even if the constants are fixed throughout the Universe we can observe, could there in some sense be other universes where they are different? This idea was outlined by a biologist, C. F. A. Pantin. He said "the properties of the material universe are uniquely suitable for the evolution of living creatures. If we could know that our universe was only one of an indefinite number with varying properties we could perhaps invoke a solution analogous to the principle of natural selection, that only in certain universes, which happen to include ours, are the conditions suitable for the existence of life, and unless that condition is fulfilled there will be no observers to note the fact."

If someone walks into a clothing shop and buys a suit that is a perfect fit to his or her body, there are two possibilities. Either the tailors who work in that shop have carefully measured that person's body and made a suit to fit it—bespoke tailoring—or the shop has such a large range of clothing available, in all shapes and sizes, that the person in question has been fitted out from stock, off the peg. The idea that the Universe is in some way constructed for our benefit, or at least designed as a fit home for intelligence, corresponds to the first possibility. In many ways, the second alternative is more attractive; but it requires the existence of a vast array of alternative universes from which we have "chosen" by the fact of our existence. In this picture, there are myriads of other worlds in which the laws of physics and the constants of nature do differ, a little or a lot, from those we know. In most of the universes, life—certainly intelligent life—does not exist. Any universe in which our kind of intelligent life can arise must look rather like our Universe, since without the familiar coincidences and constants that life would not be there. We believe our Universe to be special because we inhabit it. But that does not mean that it is special in any deeper sense of the word.

A useful analogy is with a lottery. Suppose a million lottery tickets are sold, and then one number out of that million is selected. The holder of that number wins the prize, so that number seems special. But in a deeper sense it is no more special than any of the other numbers in the lottery. By the nature of the lottery, *somebody* must win, and each of the numbers has an equal chance of winning. It is only after the event that one number gains a special status. The holder may feel lucky as a result; but somebody *had* to get lucky!

Maybe the world is like that. There may be a multitude of universes that all start out sterile. Intelligence appears in some (or perhaps only one) of those universes as a result of the accumulation of random coincidences ("luck"). But there is no meaning to the coincidences, and that universe stands out from the rest as special only with hindsight, once intelligence has appeared to wonder over its own origins.

The Quantum Realities

The key to this off-the-peg approach to the anthropic coincidences is that there must indeed be a variety of universes, an ensemble, to choose from. Science fiction readers will already be familiar with the concept*; it happens that it also has a perfectly respectable scientific basis, in quantum theory.

Quantum physics is all about probabilities. In an honest lottery, every ticket has an equal chance of winning; but the quantum world is not like that. The position an electron will be found in when we make a measurement on an atom, for example, depends on quantum probability. There is a very high probability that the electron will be found in one of the energy states corresponding to the "steps" on that particular atom's energy ladder, and a very small probability that it will be found somewhere else entirely, not even connected to the atom where we expect to find it. Rather as if the lottery is rigged, so that any number ending in 9 has a good chance of winning, while any other number has only a low probability of coming out of the hat. When we make a measurement, we may say that we have located the position of the electron, or at least observed which energy step it is sitting on. But that does not mean that another measurement will give us the same result. In the quantum world, as soon as we stop looking at an electron, it dissolves into a mist of probabilities, called a superposition of quantum states. It is like the lottery before the winning number is drawn. Making the same observation again will give a new answer, as if we put the winning number back into the hat and made a second draw—probably getting another number ending in 9, perhaps a different number altogether, possibly (but not certainly) the same number as before. The act of measurement forces the electron to choose among the possible states and to take a solid identity.

*Olaf Stapledon's classic *Star Maker*, for example, although first published back in 1937, contains some surprisingly modern-sounding descriptions of universes with different physical laws, and even different numbers of dimensions of both space and time.

But each identification is the result of a separate, independent coalescence, what is known as a "collapse of the wave function."

This is the basis of the standard interpretation of quantum mechanics, called the Copenhagen interpretation. It works very well if you want to apply the quantum rules to designing a laser, say, or to calculating how atoms join together to make molecules. But it falls down completely if you try to imagine a way to describe the entire Universe in accordance with quantum theory. Any quantum system, according to the Copenhagen interpretation, exists as a superposition of states, a nebulous array of probabilities, unless and until it is observed from outside. But what is there "outside" to observe the Universe and to collapse its wave function? This puzzle has led some cosmologists to embrace a different interpretation of quantum mechanics, which is called the Many Worlds interpretation.

Many Worlds quantum mechanics has long been regarded with suspicion by physicists. Where Copenhagen theory says that none of the quantum options has any reality unless it is observed, Many Worlds theory says that all quantum possibilities really do exist, each in its own space and time, and that each measurement we make simply identifies which branch of the multiuniverse we are in. There is a separate universe, in this picture, for each possible energy level that our electron might inhabit. When we measure the atom, we find the electron in the one energy state that corresponds to the universe we live in; but our doppelgängers in the universe next door may simultaneously be making the same measurement and getting a different answer.*

---

*"Next door" and "simultaneously" are rather difficult concepts in this connection; it's better to think of all the different worlds being at right angles to each other, perpendicular rather than parallel. This is discussed by John Gribbin in *In Search of Schrödinger's Cat.*

Both the Copenhagen and the Many Worlds interpretations give exactly the same "answers" when applied to practical problems like the design of a laser. It is simply a matter of philosophical preference which one you choose to work with, at that level. Most physicists and engineers don't care about the philosophy and use a set of rules derived from the Copenhagen interpretation, which came along first (as John Polkinghorne puts it, "the average quantum mechanic is no more philosophical than the average motor mechanic"). But there is one problem that the Many Worlds theory can deal with but the Copenhagen interpretation cannot. This happens to be important enough to give the Many Worlds theory the edge—it is the problem of providing a quantum mechanical description of the entire Universe.

In the Copenhagen theory, the Universe cannot exist, in the everyday sense of the word, unless something outside the Universe measures it and collapses the wave function. But in the Many Worlds theory all possible universes exist. The differences between some universes are trivial—here an electron is on the first step of an energy ladder, there it is on the second step. The differences between others are more extreme—a change by a factor of 10^{10}, perhaps, in the strength of the gravitational force. But each universe is "real" in the everyday sense of the word, and it is no longer a surprise to find that the Universe we inhabit fits us exactly, since we have simply selected it, off the peg, by our existence. Among the many worlds, we wear the one that fits.

But this is not the only way in which to imagine an ensemble of universes from which, by our existence, we "choose" to live in one suitable for life. If the Universe is infinite, then anything that can possibly happen may happen, or may have happened, or may be happening somewhere in that infinity. There could be a world, somewhere in the infinite Universe, where you wrote this book, and we are the readers; a world where Virginia is still a colony of England (and one where England is a colony of Virginia); and so on. These worlds will be separated from us, in an infinite Universe, only by distance, not by any mind-numbing extra dimensions of space and time at right angles to each other. Alas, though, we cannot view them, because their light has not yet had time to reach us.

It seems, at first sight, even more like an outrageous offshoot of science fiction than the Many Worlds theory. But it is now being taken very seriously, as an offshoot not of SF but of the inflationary theory of cosmology.

Inflation in a Nutshell

"Inflation" is a generic name given to a set of theories that attempts to explain how the entire presently observable Universe could have homogenised itself while expanding from an initial superdense state at a time corresponding to 10^{-43} seconds (the Planck time) after the moment of creation to the state, less than a millisecond later, when its density was roughly that of an atomic nucleus. From then on, everything can be described in terms of the laws of physics we know from our experiments and observations on Earth, within the framework of the standard Big Bang theory. But what happened before that time?

Inflation describes these events in terms of the way the original symmetries between the forces of nature broke when the Universe was young. Using the best theories we have, the grand unified theories, it is possible to calculate the energy at which this should have happened, and this corresponds to a time when the age of the Universe was about 10^{-35} seconds. The simplest thing that could have happened at that time was that the forces would quietly have gone their separate ways. But that would not have provided the expansionary boost needed to smooth out any wrinkles in the fabric of spacetime and make the Universe flat. This is where inflation comes in.

The way the Universe was cooling and changing from one state to another at that time can be compared with the way water vapour cools and condenses into liquid water. This is a change of state, from vapour to liquid, in which energy is released. The comparable change in the early Universe when symmetry was broken is known as a phase transition, and also releases energy. In 1980, Alan Guth, of MIT, suggested an extension of this analogy. Sometimes, water vapour can be cooled below 100 degrees C, the temperature at which it "ought" to become liquid. The energy that ought to be released at the transition is locked up, and the supercooled system becomes increasingly unstable as it cools further, until suddenly all the vapour changes into liquid and a burst of heat is given off. Guth suggested that the equivalent process could happen in the early Universe, with the phase transition being delayed while the whole Universe supercooled, and then flashing over into the broken-symmetry state, releasing all the locked-up energy of the phase transition in the process.

During the supercooling phase, the calculations show, the Universe will be driven into a wild, exponential expansion—inflation. It will double in size roughly once every 10^{-34} seconds, which doesn't sound too impressive until you realise that this means a hundred doublings in the space of 10^{-32} seconds, enough to expand a tennis ball to the size of the entire observable Universe. At the end of inflation, the Universe was *extremely* flat, with all the wrinkles ironed out; and the burst of energy from the delayed phase transition then heated it up, establishing the familiar conditions of the standard Big Bang as the runaway expansion at last began to slow.

There are hopes—not yet fully realised—that these theories can account for the fluctuations that eventually gave rise to galaxies. The amplitude of these fluctuations (the size of the ripples perturbing the flat background Universe) is an important cosmic number that still lacks a proper explanation.

Since Guth proposed the idea of cosmic inflation, it has gone through many changes and now exists in many different forms proposed by different theorists. The basic principle is very attractive and solves many puzzles of how the Universe got to be in a Big Bang state. But there is still the puzzle of how the seed that inflated to become our Universe—a hot, dense concentration of spacetime smaller than a proton but containing all the mass-energy that became the observed Universe—came into existence at 10^{-43} seconds. One possibility is that the basic state of the universe at large, the meta-universe, is one of chaos, with some regions expanding, some contracting, some hot, and some cold. In some parts of this infinite meta-universe conditions just happen to be right for inflation to begin, and so it does. A universe is born. But the most dramatic possibility is that the infinite meta-universe is, was, and always will be in a state of inflationary expansion itself, with a temperature of about 10^{31} K and a density of about 10^{93} grams per cubic centimetre. In its 1980s incarnation, the idea stems from the work of Richard Gott at Princeton University and Andrei Linde at the P. N. Lebedev Institute in Moscow—although it has curious echoes of the steady state theory, developed by Fred Hoyle and Jayant Narlikar, which fell from favour in the 1960s. And it all derives from one of the earliest solutions discovered, in 1917, for Einstein's equations of general relativity.

Bubbles on the River of Time

"Many and strange are the universes that drift like bubbles in the foam upon the River of Time." When Arthur C. Clarke wrote those words, almost forty years ago, as the opening to a science fiction story called "The Wall of Darkness," he can have had no idea that in the late 1980s they would stand as an accurate description of modern cosmological thought. Theorists are now being led to consider the possibility that our Universe is indeed just one bubble among many in some greater meta-universe.

One way of imagining the seed of a universe like our own coming into existence as a tiny concentration of mass-energy with an "age" of 10^{-43} seconds is simply as a quantum fluctuation of the vacuum, one of those things permitted by uncertainty, like the appearance of a virtual pair of particles out of nothing at all. The idea surfaced in the early 1970s, but in its original form the trick didn't work. Such an enormously massive fluctuation could indeed occur in principle, but it would occupy a tiny volume of space, smaller than a proton, and it would, by definition, have an enormous gravitational field. The result would be an extremely rapid collapse, snuffing the embryonic universe out of existence as quickly as any pair of virtual particles produced in the vacuum today. Inflation, however, provides a way out of this dilemma. In the tiny fraction of a second it exists, inflation can set to work and blow the seed up into a full-size universe. "Full size," in this case, means something about as big as a basketball, containing as much mass-energy as the entire visible Universe today, and experiencing a big bang. All the rest follows naturally from the known laws of physics.

But why stop at one universe? If bubbles of mass-energy can appear out of nothing at all, and explode exponentially into life as fully fledged universes, shouldn't the same processes—vacuum fluctuations—be going on in the space between the stars today? And, if so, might it not be rather uncomfortable for us if a new universe popped into existence nearby? Yes, and no. Other universes might indeed be being born all the time in the vacuum of space; but if they were, there is no way that we would know about it.

The possibilities have been investigated by several researchers, among them Edward Fahri and Alan Guth. They envisage artificially creating other universes; their scientific paper on the topic is entitled "An Obstacle to Creating a Universe in the Laboratory." You don't need a lot of mass to start with; the quantum effects will provide the mass-energy of a universe for you, once the process is kicked into starting. But you do need to create conditions of very high density and a temperature equivalent to about 10^{24} K, in order for inflationary processes to do their thing. We already have the energy available to do half the trick, in the form of hydrogen bombs; the other half of the trick is to confine that energy within a very small volume (the size of an atom) which is why nobody is manufacturing universes in their basement just yet. If you *could* confine the energy, though, what you ought to get would be, according to the equations of general relativity, a black hole. Interesting in its own right, but not a new universe. Or is it?

Guth and others have shown that what happens inside the confined region depends on exactly how the pressure is applied. In many cases, the compressed region does turn out to be "only" a black hole. But there are solutions to the equations that, given the right initial conditions, do allow for the prospect of inflation. The confined region does not, however, expand back out into the Universe at large. Instead, it expands in a direction at right angles to our familiar dimensions of space and time, off into a universe of its own. Exactly the same thing will happen to any inflationary seeds created by quantum fluctuations in the vacuum of our Universe.

An analogy that is often made is to describe our expanding Universe as the skin of a balloon that is steadily increasing in size (cosmologists used to say "an inflating balloon," but we are now talking about the present-day expansion of the Universe, not its inflationary past, so we have to be careful with our choice of words). The two-dimensional skin of the balloon represents *all* of our familiar dimensions. As the balloon expands, the Universe gets bigger. Any "new" universes created within our Universe, either naturally or by someone with an H-bomb in his basement, are like little bubbles in the skin of the balloon. They pinch off from our spacetime (the skin of the balloon) and expand outwards in their own right, in their own space and their own time.

From our perspective, nothing seems to have happened. Perhaps a black hole has appeared, perhaps not. From the perspective of any observer able to withstand the extreme conditions inside the superdense region, however, things would be very different. The region would inflate exponentially, then go over into a big bang and expand more sedately. Stars, galaxies, and intelligent creatures could evolve, study their surroundings, and begin to wonder about the possibility of creating new universes in the basement (provided the laws of physics in the new universe permitted intelligence to evolve; this is by no means certain, since each universe may have its own set of laws and constants). Quantum cosmology allows the possibility of creating not just one universe but an infinite number of universes out of nothing at all. The universes may be interconnected, in some complex way, as new universes are born within, but then pinch off from, the vacuum of old universes, producing a complex, multidimensional foam. Our Universe may simply be a region of spacetime that has pinched off from another bubble. But the bubbles can never communicate with one another and might have very different properties from one another. The end of inflation is linked with the breaking of symmetry between the four forces of nature; there is nothing to say, however, that the symmetry will break in the same way in every bubble. In some bubbles, the forces will have different strengths from those in our Universe; indeed, there may be three, or five, fundamental forces, or some other number, instead of the four we know.

We are back in the worlds of Arthur Clarke. If an infinite variety of universes exists, then all things are possible. There must be infinitely many universes where gravity is too weak for life to emerge, infinitely many where gravity is too strong, and infinitely many more where something else goes wrong. But it is no puzzle that we exist, since there must also be an infinite number of bubbles in which conditions closely resemble those we see in the Universe around us, and the world is, like baby bear's porridge, "just right."

Cosmic Dragons

The anthropic principle cannot claim to be a scientific explanation in the proper sense. At best it can offer a stop-gap satisfaction of our curiosity regarding phenomena for which we cannot yet obtain a genuine physical explanation. The most powerful of these insights may be the hint that our Universe is not unique, but just one among an ensemble of universes, whatever form that ensemble might take. Andrei Linde envisages an infinite universe divided into domains in each of which the physics would be different. Most of this meta-universe would be a lifeless desert; complex evolution would occur only in "oases" where the constants—the numbers of dimensions and so on—had propitious values. Our oasis must then be at least 10 billion light-years across, because the laws of physics seem to be the same everywhere we look. But the desert regions beyond may in principle be observable in the remote future when, perhaps a thousand billion years or more from now, light from the edges of our domain has had time to reach us. This is too remote an eventuality to provide a practical way of testing the possibility—but the conceptual status of the idea is no different from the conjectures of early cosmographers about continents beyond the horizons of the then-known world. We prefer to attempt at least to sketch the outlines of the continents that might lie beyond our present horizons, rather than simply fill in the edges of the cosmological map with the legend "here be dragons." Anthropic reasoning suggests that those other worlds do indeed exist, even if we can never have direct knowledge of them.

"The most incomprehensible thing about the Universe is that it is comprehensible" is one of the best known of Einstein's sayings—it has become a cliché. He meant by it that the basic physical laws, which our brains are attuned to understand, have such broad scope that they offer a framework for interpreting not just the everyday world but even the behaviour of the remote cosmos. The physicist Eugene Wigner described this as the unreasonable effectiveness of mathematics in the physical sciences. Cosmologists start by using the physics that is validated locally, and apply this, with simplifying assumptions, to probe the workings of the Universe at large. These simple rules seem to work as a description of the Universe. There seems no reason why the Universe should be so structured that this approach permits any real progress—why the physics we study in the laboratory on Earth applies also in quasars billions of light-years away and in the early stages of the Big Bang. Unless, perhaps, there *is* some link between the simplicity of the Universe and its suitability as a home for intelligent life. To say that we would not be here if things were otherwise, however, need not quench our curiosity and surprise at finding that the world is as it is.

The Philosophy of Cosmology

What, then, is the physical status of anthropic reasoning— anthropic cosmology—today? Some people adopt the dismissive attitude that anthropic reasoning cannot offer scientific explanations in the proper sense. At best, they say, it can give a stop-gap satisfaction to our curiosity regarding phenomena for which we have as yet no genuine physical explanation. The world would indeed be very different if the relative strength of the nuclear and electromagnetic interactions were somewhat altered, but one still hopes for a unified physical theory that predicts the actual constants or relates them to one another. A little over a century ago, theorists might have imagined varying the electrical and magnetic forces and the speed of light—before the work of James Clerk Maxwell showed how these were interconnected. By extension, a more comprehensive theory may eventually relate all the fundamental forces. Most theorists indeed hope that the constants of nature will not forever have to be treated as numbers derived from experiments, but will be related by a unified theory. They will then be mathematically calculable, just as a circle's circumference can be calculated (rather more easily!) from its diameter.

A hostile view of the anthropic principle comes from Heinz Pagels's book, *Perfect Symmetry*: written in 1985:

Physicists and cosmologists who appeal to anthropic reasoning seem to me to be gratuitously abandoning the successful program of conventional physical science of understanding the quantitative properties of our universe on the basis of universal physical laws. Perhaps their exasperation and frustration ... has gotten the better of them. ... The influence of the anthropic principle on the development of contemporary cosmological models has been sterile. It has explained nothing, and it has even had a negative influence, as evidenced by the fact that the values of certain constants, such as the ratio of photons to nuclear particles, for which anthropic reasoning was once invoked as an explanation can now be explained by new physical laws. ... I would opt for rejecting the anthropic principle as needless clutter in the conceptual repertoire of science.

We think this is going too far in disparaging anthropic reasoning. After all, in its weak form, little more is involved than the routine attitude of an experimenter who takes account of the limitations of laboratory techniques and equipment.

The case for the defence, however, is presented with baroque elaboration by John Barrow and Frank Tipler in their massive book *The Anthropic Cosmological Principle*. Without going all the way with them, we agree that the principle does deserve serious attention. Its eventual status will depend on what the laws of nature are really like. If some final unified theory yields *unique* numbers for all the constants, then it may be inconceivable to envisage a different kind of universe. But if the basic laws turn out to involve some random or statistical element, then the idea of an ensemble of universes, outlined in this chapter, could be put on a serious footing. It could then indeed be natural selection, not mere accident, that our Universe (that is, the part of spacetime we can observe) has the particular values of the physical constants that we measure.

The "weak" anthropic principle—the realisation that the existence of observers such as ourselves imposes some selection effects on what we see around us—is almost banal. Any more pretentious role for anthropic reasoning is controversial, and depends on the true nature of the laws of physics. To quote Steven Weinberg, from a 1984 BBC broadcast, "I certainly wouldn't give up attempts to make the anthropic principle unnecessary by finding a theoretical basis for the values of all the constants. It's worth trying, and we have to assume that we shall succeed, otherwise we surely shall fail."

So perhaps it is best, if they are to retain their scientific motivation, that theoretical physicists should not take the strong anthropic principle, the idea that the Universe is tailor-made for man, *too* seriously. If there is a unique "theory of everything," then there is certainly a sense in which the laws of physics could not have been otherwise. We would then have to accept it as genuinely coincidental, or even providential, that the constants determined by high-energy physics happen to lie in the narrowly restricted range that allows complexity and consciousness to evolve in the low-energy world we inhabit. The intricacy implicit in these unique laws may astonish us, but our reaction would be no less subjective than a mathematician's surprise at the rich intellectual structures that can stem from simple axioms. Everything would be a consequence of unique laws. But that need not spell the end of scientific investigation of our surroundings.

The End of Physics?

Physicists sometimes talk of a theory of everything, or TOE, as providing "the end of physics," in the form of a single package of equations to describe the Universe and all it contains. But that would not really be the end of science, or even put all physicists out of work overnight. No set of equations explains why there *is* a universe. To quote Stephen Hawking, "What is it that breathes fire into the equations? Why does the Universe go to all the bother of existing?" In any case, most of the challenging questions we ask about the natural world, on the astronomical as well as the terrestrial scale (other than those involving initial conditions or "origins") involve "old-fashioned" atomic and nuclear physics. The subnuclear world and the uncertainties of high-energy physics are generally irrelevant to larger-scale phenomena, just as the atomic structure of liquids provides no practical clues to the still unexplained complexities of turbulent flows in the air and the oceans. We can in principle write down the equations (essentially the ones derived by Erwin Schrödinger in the 1920s) governing the system—but we cannot *solve* the equations even for a typical single molecule, let alone for any larger system. Nor, even if we could solve the equations, would we have enough accurate information about the starting conditions (the position and velocity of every molecule) to permit accurate predictions. Even if we are "reductionists," believing that all phenomena can be reduced to physical fundamentals, this does not permit us to be "constructionists," in the sense of deriving a full understanding of complex systems from their atomic constituents. Sciences will always be in a hierarchy, where each level of structure entails new irreducible concepts.

Suppose you were unfamiliar with the game of chess. Just by watching a game being played, you could infer what the rules were. The physicist, likewise, finds patterns in the natural world, and learns what dynamics and transformations govern its basic elements. But in chess, learning how the pieces move is just a trivial preliminary to the absorbing progression from novice to grand master. The whole point and interest of the game lies in exploring the complexity implicit in a few deceptively simple rules. Uncovering a TOE would do no more (and probably much less) than the equivalent of putting us in the status of a novice chess player who has just opened the book of rules—and knowing all the rules of chess does not permit even an expert to predict the outcome of a match between two grand masters, let alone every move played.

Biologists aim to delineate 3.5 billion years of the evolutionary history of life on Earth—to learn how, in Darwin's words, "whilst this planet has gone cycling on according to the fixed law of gravity, from so simple a beginning endless forms most beautiful and most wonderful have been, and are being evolved." Astrophysicists and cosmologists aim to set the Earth, and the entire Solar System, in a broader context of cosmic evolution. Progress is meagre, but what is astonishing is that there is any progress at all. We have 10 billion years, or more, of the past history of the universe to ponder, and far more than 10 billion years of Universal future to contemplate. The task of the evolutionary biologist pales by comparison—yet nobody would seriously claim that we understand every facet of life on Earth. Breaking the problem down into smaller pieces, we can contemplate the possibility of finding a "solution" to the puzzle of galaxy formation, for example, but we have no real idea where to start in finding out how life began. We cannot say whether life is common or rare in the Universe, nor whether it is confined to a single planet. There is ample work yet for physicists, and others!

Astronomers would confidently argue that planetary systems, potential abodes for life, are widespread in our Galaxy (and presumably in every other). But, even given the right environment, what are the odds on life getting established and evolving to an "interesting" stage? We claim no expertise on this subject, but have the impression that there is no consensus even among the experts. The chance *could* be high [and a "SETI" search is certainly, given the colossal pay-off if it succeeds, a worthwhile gamble]; but it could on the other hand, be so low that life is very rare. If the odds were, say, 1 in 10^{20} or less, there might be no other life within the entire observed universe.

Brandon Carter has advanced an *anthropic* argument suggesting that life may indeed be rare. He notes an interesting biological coincidence, that the Sun is half way through its life, and it has taken that long for intelligence to evolve on Earth. Stellar lifetimes are a straightforward consequence of the physical laws and constants; biological evolution, on the other hand, is an immensely complex multistage process. There seems no conceivable reason why these times should be closely comparable. Carter conjectures, therefore, that the *typical* time taken for biological evolution is *much greater* than stellar ages. Evolution, he claims, must have been especially rapid on Earth.* On this hypothesis, one can make the definite prediction that intelligent life would be rare in the Universe. In typical cases, even if biological evolution got started, it would not get very far before the host star died.

Even if life is rare, however, and even if we live in just one bubble universe among many in the foam on the river of time, there may be infinite time ahead of us, and infinite space in which to develop. In this cosmic perspective, we may still be near the "simple beginning" of the evolutionary process—certainly not its culmination. So living things may eventually become an important part of the Universe, modifying the astro-

---

*There are other possible reasons why our planet should be unusual in this regard. To mention just one, the Earth has an unusually large Moon, making the Earth-Moon system almost a double planet. Could the unusually large tides raised on Earth as a result have influenced the evolution of life?

nomical environment in the way that mankind is already modifying the terrestrial environment— although we might hope that our descendants would take a little more care with their modifications to stars and galaxies than we have taken with our modifications to the Earth. It is sometimes argued that if life is a rare accident, it would be an irrelevant fluke in a mindless and hostile cosmos. But we take the opposite viewpoint: if there is no life elsewhere, the Earth acquires universal significance as the spark with the unique potential to spread life and consciousness through the cosmos. We should regard present life on Earth as the beginning of a process with billions of years, and perhaps a literally infinite timespan, still to run—the greening of our Galaxy and beyond by forms of life and intelligence (not necessarily all organic) seeded from Earth. To snuff out our biosphere would quench the evolutionary process when it had barely begun to realise its limitless potential. We are back in the realms of Olaf Stapledon—an entirely appropriate note on which to end our discussion of the interrelationship between humankind and the cosmos, since there is one key ingredient in science, which is highlighted most effectively by the development of anthropic cosmology—a sense of wonder.

Further Reading

Barrow, John, and Frank Tipler. *The Anthropic Cosmological Principle*. New York and London: Oxford University Press, 1986.

Bartusiak, Marcia. *Thursday's Universe*. New York: Times Books, 1986.

Bear, Greg. *Blood Music*. New York: Ace Books, 1986.

Chandrasekhar, Subrahmanyon. *The Mathematical Theory of Black Holes*. Oxford: Oxford University Press, 1982.

Davies, Paul. *The Search for Gravity Waves*. New York and London: Cambridge University Press, 1980.

Davies, Paul, and Julian Brown (eds.). *Superstrings*. New York and London: Cambridge University Press, 1988.

Dawkins, Richard. *The Blind Watchmaker*. New York and London: Penguin and Norton, 1987.

Gribbin, John. *In Search of Schrödinger's Cat*. New York: Bantam; London: Corgi, 1984.

———. *In Search of the Big Bang*. New York: Bantam; London: Corgi, 1986.

———. *The Omega Point*. New York: Bantam; London: Corgi, 1988.

Hawking, Stephen. *A Brief History of Time*. New York and London: Bantam, 1988.

Henderson, Lawrence. *The Fitness of the Environment*. Cambridge, Mass.: Harvard University Press, reprinted 1970.

Hoyle, Fred. *Galaxies, Nuclei, and Quasars*. London: Heinemann, 1965.

Hubble, Edwin. *The Realm of the Nebulae*. New York: Dover, 1958.

Kaufmann, William. *Universe* (second ed.). New York and Oxford: Freeman, 1984.

Narlikar, Jayant. *The Primeval Universe*. New York and London: Oxford University Press, 1988.

Pagels, Heinz. *Perfect Symmetry*. New York: Bantam, 1985.

Patin, C. F. A. *The Relations Between the Sciences*. Cambridge: Cambridge University Press, 1968.

Polkinghorne, John C. *Particle Play*. New York: Freeman, 1979.

———. *Quantum World*. Princeton and London: Princeton University Press and Penguin, 1985.

Rees, Martin. *Quasars, Black Holes and Galaxies*. Freeman, New York & Oxford).

Shu, Frank. *The Physical Universe*. Mill Valley, Calif.: University Science Books, 1982.

Stapledon, Olaf. *Star Maker*. Los Angeles: J. P. Tarcher, reprinted 1987.

Weinberg, Steven. *The First Three Minutes*. New York: Bantam; London: Flamingo, 1976.

Will, Clifford. *Was Einstein Right?* New York: Basic Books, 1986.

INDEX

★

About the Authors

Dr. John Gribbin, science writer and cosmologist, is the author of many nonfiction books, including *In Search of the Big Bang, In Search of the Double Helix,* the best-selling *In Search of Schrödinger's Cat,* and *The Omega Point,* which was nominated for Britain's top science book award in 1988. His first solo science fiction novel, *Father to the Man,* has recently been published in the U.S. A frequent guest on science programs on BBC Radio, Gribbin holds a doctorate in astrophysics from Cambridge University and lives in East Sussex, England.

While still a graduate student, Dr. Martin Rees predicted that structures inside quasars (then newly discovered) would appear to move much faster than light. This was confirmed several years later, and he has continued to be on the forefront in the study of cosmology, galaxies, black holes, and space science. Elected Plumian Professor of Astronomy and Experimental Philosophy at Cambridge when he was only 30, Rees continues to be interested in the broader philosophical implications of his research and in conveying the fascination of the latest scientific ideas to a wide public. He has written numerous general articles, appeared frequently on radio and television, and given many public lectures, especially in the United States.

Rees lives with his wife, a social anthropologist, in an old farmhouse near Cambridge.